CAMBRIDGE LIBRARY COLLECTION

Books of enduring scholarly value

Physical Sciences

From ancient times, humans have tried to understand the workings of the world around them. The roots of modern physical science go back to the very earliest mechanical devices such as levers and rollers, the mixing of paints and dyes, and the importance of the heavenly bodies in early religious observance and navigation. The physical sciences as we know them today began to emerge as independent academic subjects during the early modern period, in the work of Newton and other 'natural philosophers', and numerous sub-disciplines developed during the centuries that followed. This part of the Cambridge Library Collection is devoted to landmark publications in this area which will be of interest to historians of science concerned with individual scientists, particular discoveries, and advances in scientific method, or with the establishment and development of scientific institutions around the world.

Six Months in Ascension

Six Months in Ascension, first published in 1878, contains an account by Isabel Sarah B. Gill of the 1877 scientific expedition to the island of Ascension, in the South Atlantic, undertaken to measure the distance of the sun from the earth by observing the opposition of the planet Mars. The expedition, funded by the Royal Astronomical Society, was led by Isabel's husband, the astronomer David Gill, with a heliometer and other scientific instruments provided by Lord Lindsay. Isabel accompanied the expedition as her husband's companion. Her account offers personal details and stories omitted from the scientific reports on the expedition written by her husband and colleagues and it contains beautiful descriptions of the island of Ascension. The book offers a rare view of the personal, practical and behind-the-scenes side of a nineteenth-century scientific expedition and provides a fascinating insight into the gender roles of learned Victorian society.

Cambridge University Press has long been a pioneer in the reissuing of out-of-print titles from its own backlist, producing digital reprints of books that are still sought after by scholars and students but could not be reprinted economically using traditional technology. The Cambridge Library Collection extends this activity to a wider range of books which are still of importance to researchers and professionals, either for the source material they contain, or as landmarks in the history of their academic discipline.

Drawing from the world-renowned collections in the Cambridge University Library, and guided by the advice of experts in each subject area, Cambridge University Press is using state-of-the-art scanning machines in its own Printing House to capture the content of each book selected for inclusion. The files are processed to give a consistently clear, crisp image, and the books finished to the high quality standard for which the Press is recognised around the world. The latest print-on-demand technology ensures that the books will remain available indefinitely, and that orders for single or multiple copies can quickly be supplied.

The Cambridge Library Collection will bring back to life books of enduring scholarly value (including out-of-copyright works originally issued by other publishers) across a wide range of disciplines in the humanities and social sciences and in science and technology.

Six Months in Ascension

An Unscientific Account of a Scientific Expedition

ISOBEL SARAH BLACK GILL
DAVID GILL

CAMBRIDGE
UNIVERSITY PRESS

CAMBRIDGE UNIVERSITY PRESS

Cambridge, New York, Melbourne, Madrid, Cape Town, Singapore,
São Paolo, Delhi, Dubai, Tokyo

Published in the United States of America by Cambridge University Press, New York

www.cambridge.org
Information on this title: www.cambridge.org/9781108014281

© in this compilation Cambridge University Press 2010

This edition first published 1878
This digitally printed version 2010

ISBN 978-1-108-01428-1 Paperback

SIX MONTHS

IN

ASCENSION

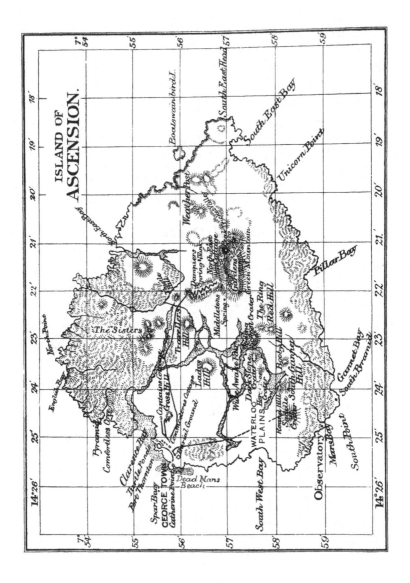

ISLAND OF
ASCENSION.

SIX MONTHS

IN

ASCENSION

𝔄𝔫 𝔘𝔫𝔰𝔠𝔦𝔢𝔫𝔱𝔦𝔣𝔦𝔠 𝔄𝔠𝔠𝔬𝔲𝔫𝔱 𝔬𝔣 𝔞 𝔖𝔠𝔦𝔢𝔫𝔱𝔦𝔣𝔦𝔠 𝔈𝔵𝔭𝔢𝔡𝔦𝔱𝔦𝔬𝔫

By MRS. GILL.

WITH A MAP.

LONDON :

JOHN MURRAY, ALBEMARLE STREET.

1878.

INTRODUCTION.

—◆—

A SCIENTIFIC expedition may be said to have two histories. The one treats of the special object of the expedition, the other of the personal adventures of those concerned in it. It is only the former which finds permanent record in the Transactions of scientific societies : the other too often remains unwritten.

For many reasons I think this is a matter of regret. Mere details of observations are never looked at, except by a very limited number of specialists; to the general public such details are meaningless as well as inaccessible ; whilst the ordinary student usually accepts the result merely as he finds it quoted in some standard work or text-book.

It is not because popular accounts of such expeditions do not interest a sufficient circle of readers that they have not been more frequently written, but rather, I think, because the faculties of original research and popular exposition are seldom united in the same individual. Besides this, I have found in my own experience, that on such expeditions there is so much actual work to be done, and the hours are so completely filled with it, that there is neither time nor inclination to write a diary. Thus, before the story can be committed to writing, it has lost its crispness—the interest has faded, and, from treacherous memory, incident is wanting to complete the narrative. On my expedition to Ascension last year, however, I had the good fortune to be accompanied by my wife, who found much pleasure and interest in making a daily record of our life and work there. This little book, compiled from her journal, she now lays before the public with much diffidence. It is an honest endeavour to tell a true story, and add somewhat to a neglected class of literature; as such, she hopes that the faults incidental to a first work will meet with lenient judgment.

The story can boast of no stirring interest, no thrilling adventures by land or sea. It must derive its interest chiefly from its truthfulness as a record of an attempt to solve a great problem, viz., *the distance of the Earth from the Sun.* The nature of this problem my wife explains in Chapter I. sufficiently, I think, to make it of interest to the many that would gladly learn something of the history of scientific progress, but who are often deterred from so doing by the minute details and heavy technicalities with which its every step is necessarily encumbered.

To the best of her ability she has given *one* side of the history of *one* step.

It now remains for me to preface her chronicle by a brief outline of the labours of those who have worked before in the same field of research.

The first attempt to measure the Sun's distance was made as follows.

In the diagram suppose s the Sun, m the Moon, and e an observer on the Earth. When the angle s m e is a right angle, the Moon will be exactly half full. If it is less than a right angle the Moon will appear less than half illuminated to an observer at e, and *vice versâ.* Hence, if the angle s e m is accurately measured at the instant when the Moon is half full, the proportions of the triangle s e m will be known; because the angle s m e being a right angle, and the angle s e m having been determined by measurement, one side and two angles are known; the triangle s e m can therefore be drawn on paper, or its proportions may be computed mathematically.

Since E M represents the distance of the Earth from the Moon at the instant of observation, the proportion of this distance to that of the Sun (represented by the line E s) is determined.

In this way Aristarchus of Samos concluded that the Sun was nineteen times more distant than the Moon. This distance we now know to be more than twenty times too small—and the reason of his failure was twofold. 1st. Because, from the irregularity of the Moon's surface, it is almost impossible to estimate when she is exactly half full; 2nd. Because his means of measuring the angle of s E M were rude and imperfect. The result of Aristarchus was however adopted by astronomers till the time of Kepler.

About the year 1620, from the observations of Tycho Brahé on the planet Mars, Kepler concluded that the distance of the Sun must amount at least to 1800 diameters of the Earth, upon which he was upbraided by his friend Crüger " for removing the Sun to such a huge distance."

The first approximation to a true determination was the result of an expedition organized by the French astronomer, Cassini. Richer had been sent to South America by the French Academy. On the 1st of October, 1672, the planet Mars approached very close to a bright star (ψ Aquarii), and, from observations made by Richer in Cayenne, and by Picard and Romer in France, Cassini concluded that the Sun's distance must be at least 86 millions of miles.

The results were, however, liable to very considerable uncertainty. This was due to the imperfect instruments of the time, by means of which it was hardly possible to measure angles with the required accuracy.

The first satisfactory step in advance is due to the Abbé de la Caille, whose celebrated expedition to the Cape of Good Hope took place in the year 1740. There he made a large number of observations of Mars, and from these, compared with corresponding observations in the Northern Hemisphere, deduced $81\frac{6}{10}$ millions of miles for the Sun's distance. He afterwards combined this result with similar observations on the planet Venus, and arrived at $81\frac{5}{10}$ millions of miles as a final result.

In the meantime, however, the English astronomer Halley had gone to St. Helena, where he observed the Transit of Mercury on the 28th of October, 1677 (see Chap. III. p. 33), and this suggested to him the method of determining the Sun's distance by the Transits of Venus which would take place in the years 1761 and 1769.

In two remarkable memoirs presented to the Royal

Society of London in 1691 and 1716, he pointed
out the great advantages of this method, and urged
upon astronomers the necessity of providing for the
complete and accurate observation of these phenomena.
Accordingly, in 1761, England sent Maskelyne to
St. Helena, and Mason and Dixon were dispatched to
Sumatra. The two latter astronomers were so delayed
by the way, that, fearing they would not reach the
appointed station in time, they decided to remain at
the Cape of Good Hope; and the decision proved a
fortunate one.

The St. Petersburg Academy of Sciences sent
Chappe to Tobolsk, and Rumowski to Selinghinsk
near Lake Baikil in Siberia.

The French sent Pingré to the Island of Rodriguez,
and Le Gentil should have observed at Pondicherry.

Poor Le Gentil! He duly reached Mauritius on
the 10th of July, 1760—nearly a year before the
Transit. War having meanwhile broken out between
France and England, he was unable to reach Pondicherry; so he resolved to go to the Island of
Rodriguez instead, to join Pingré, who was already
there. When on the point of starting for Rodriguez,
he learned that a French frigate was about to sail from
Mauritius for the coast of Coromandel. Le Gentil
decided to avail himself of the opportunity thus

offered to reach the point chosen by the Academy ; but one delay after another occurred, and it was not until the middle of March that he sailed again from Mauritius. There was not much time to be lost, for the Transit would occur on the 6th of June. Detained by frequent calms he did not reach the coast of Malabar till the 24th of May. Still there might have been time enough to prepare for the observation, had not the commander of the frigate learned that the English were masters of Mahé and Pondicherry. His only chance to escape capture was to make off as quickly as possible. This he did ; steering a course for Mauritius again, to Le Gentil's utmost despair.

The 6th of June arrived. The sky was gloriously clear. From the deck of the vessel Le Gentil made the best observations he could ; but, from so unsteady a platform, they could be of little value to science.

Other observers had better fortune ; but the results when computed proved far from satisfactory. Different astronomers obtained results varying from $81\frac{1}{2}$ to $96\frac{1}{2}$ millions of miles ; not because there was any error in the method, but because the observations were discordant.

The difficulties of actual observation proved to be far greater than had been anticipated. Instead of precise phenomena at contact, only a gradual merging, or a gradual separation of the limbs of Venus and the Sun could be observed. Thus different interpretations could be put upon the language of the observers; and, according to the interpretations, so was the result.

But the failure of the Transit of 1761 only urged to new effort for that of 1769. Astronomers throughout the world felt that if the opportunity, which would then occur, was lost, another so favourable for determining the Sun's Distance would not occur again for 105 years. Accordingly, the most strenuous exertions were made to provide for its proper observation. Encke has well said, " Whatever may be the future judgment as to the actual issue, posterity will never be able to reproach either the astronomers or the governments of that period with having neglected to call sufficiently careful attention to the more important points, or with having failed to further and support scientific efforts with sufficient readiness."

It would occupy too much space to follow the adventures of all the observers, but some of them cannot be passed over without mention.

The French astronomer, Le Gentil—whose endeavours to observe the Transit of Venus in 1761 were defeated in the way already described—had no sooner returned to Mauritius than he set out again for Pondicherry, determined to wait there, for eight years, till the next Transit of Venus.

The eventful 3rd of June, 1769, at last arrived.

The morning was fine, and everything promised a happy issue. But, just as the critical moment approached, an unfortunate cloud eclipsed the Sun, a torrent of rain descended, *and the fruit of eight years' waiting was lost.* Le Gentil had profitably employed his time in studying the astronomy of the Brahmins, so his eight years in Pondicherry had been well occupied; but the agony of disappointment he must have felt at the defeat of his noble endeavours cannot but enlist the sympathy of all who know his story.

It was intended that the French astronomer Chappe, who in 1761 had observed the Transit of Venus at Tobolsk, should now observe it at the Solomon Islands. The Spanish Government, however, refused the necessary permission, but offered to convey him, along with two Spanish observers, to Mexico or California by a Spanish fleet then about to sail for South America.

This offer was accepted. Chappe, with his Spanish colleagues, selected Cape Lucas, in California, and there observed the Transit successfully. But he did not live to tell the tale at home. The plague visited the district, and Chappe was one of its first victims. Three days after the Transit, he was attacked by the malady, but had partly recovered, when his love of science led him to commit an imprudence which brought him to the grave. Despite his feeble condition, he passed the night of the 18th of June in observing an eclipse of the moon. A relapse ensued, and on the 1st of August he died. A short time before his death he said to his friends: "I know I have but a short time to live, but I have fulfilled my mission, and I die content."

The English Government took a bolder course, and did not wait for the permission of the Spaniards to visit the South Seas. The celebrated Captain Cook in command of the *Endeavour*, with Green (a pupil of Bradley) for astronomer, and Solander (a pupil of Linné) for naturalist, sailed on the 22nd of September, 1768, for an unknown destination. The result was one of the most brilliant and successful scientific expeditions ever undertaken. The Transit of Venus was successfully observed by Green and Cook at Tahiti, one of the Sandwich Islands; and much valuable work in connection with Natural History, Terrestrial Magnetism, and Hydrography, was accomplished.

Another English expedition was sent to Hudson's Bay. There Dymock and Wales, after encountering a good many hardships, successfully observed the Transit; and their observations acquired a special value, because it unfortunately happened that those made at the other important northern station were liable to considerable suspicion.

Father Hell, with his assistant, Father Sainovicz, was invited by the King of Denmark to observe the Transit of Venus in his dominions. In the month of June, 1768, they left Copenhagen, accompanied by Borgrewing, a Danish observer. They reached Wardoehuus, in the north of Lapland, on the 11th of October, 1768. Here they passed the winter, and duly observed the Transit of Venus in 1769. But numerous circumstances tended to throw suspicion on Hell's observations. In the first place, without any sufficient reason, he suppressed his observations for nine entire months, and many eminent astronomers did not hesitate to accuse him of having fabricated or changed them.

In 1834, his original papers were presented to the Vienna Observatory by Baron Münch-Bellinghausen, into whose hands they had come through the death of his uncle, Baron von Penkler, an intimate friend and patron of Father Hell.

Professor Littrow's investigation of these papers led to the discovery of Father Hell's original note-book for the days June 2—4, 1769.* "These notes fully

* I quote Dr. Gould's account in his admirable introduction to the U. S. Naval Astronomical Expedition to Chili.

corroborate and justify previous suspicions. The chief figures, especially the times of entrance upon the solar disc, had been for the most part erased, and with a darker coloured ink. Two passages, the one relating to the observations of Sainovicz, the other to those of Borgrewing, had been so thoroughly obliterated, that Professor Littrow was only able to conjecture the three first letters of the one and the first and last letter of the other. From an investigation of such figures as remained legible and unaltered, he succeeded in finding one observation of the Ingress by Borgrewing, and one of Egress by Hell, upon which reliance appears warrantable."

"Although in reply to Lalande, Father Hell had publicly offered to exhibit the original note-book, free from erasures, and giving observations just as finally published by him, Littrow found both clear and undefaced documents containing the quantities as prepared for publication, and this note-book, which was as manifestly not designed for press. It contains remarks, notes, and comments in chronological order; the handwriting is unequal and frequently changing, observations never made public are here noted down, together with many jottings and memoranda which could not have been intended for the public. The important observations were chiefly obliterated with great care and thoroughness, as were also sundry remarks concerning them. There can be no doubt that the evidence is sufficient to establish this note-book as being the identical one used at Wardoehuus, and that this establishment of identity discredits the published observations and the truthfulness of Father Hell, but provides few new figures upon which reliance may be placed."

Encke also found another proof of Father Hell's dishonesty. An eclipse of the Sun occurred soon after the Transit of Venus of 1769, and afforded an excellent means of checking the longitudes of the stations. Father Hell observed this eclipse, and fortunately did not change the original record in his note-book. The time he published, however, differed from the time he recorded by 41 seconds; for, in his desire to publish better observations than he knew how to make, he altered his observation to agree with his computation, which proved to have been founded on erroneous elements. His original record was afterwards found to be a good observation.

Such revelations must throw the gravest doubt upon *all* the observations of Father Hell, and give to the observations of Dymock and Wales an exceptional importance.

Although the results obtained for the distance of the Sun from the Transit of 1769 did not differ so widely as those of 1761, still the agreement was by no means satisfactory.

Thus from the numerous results may be quoted the following

		Sun's Distance	
Lalande	obtained	$96\frac{2}{10}$	millions of miles.
Father Hell	,,	$93\frac{9}{10}$	,,
Hornsby	,,	$93\frac{1}{10}$	,,
Euler	,,	$92\frac{7}{10}$	,,
Pingré	,,	$92\frac{1}{10}$	,,
Laplace	,,	$92\frac{8}{10}$	,,

In 1835 Encke published his famous discussion of the Transits of Venus of 1761 and 1769.   He found

From the Transit of 1761		$95\frac{8}{10}$	millions of miles.
,,	,, 1769	$95\frac{2}{10}$	,,
From both Transits combined		$95\frac{28}{100}$	,,

Encke's discussion met with the general approval of astronomers at the time, and for many years was accepted as the standard determination of the Sun's distance.

In 1832, Henderson observed, at the Cape of Good Hope, the favourable Opposition of Mars of that year, and his observations, combined with similar ones at Greenwich, Cambridge and Altona in the Northern Hemisphere, gave

$$90\frac{6}{10} \text{ millions of miles}$$

for the Sun's distance.   The results were, however,

not very accordant, and were not generally accepted as satisfactory.

In 1847 Professor Gerling proposed the observation of the planet Venus, at observatories in the Northern and Southern hemispheres, as a good means of determining the Sun's distance. He argued that as Venus approaches nearer to the Earth than Mars, she presents a more favourable opportunity for determining parallax. He also contended that the delicate and faint crescent form of Venus (like a very young moon when the planet is near conjunction) formed a telescopic object capable of the most accurate measurement.

Professor Gerling's idea, however, did not assume practical shape till it was taken up by a zealous observer, Lieut. Gilles of the United States Navy. He applied to his chief to ask from Congress a grant of 1000*l*. for the expenses of an expedition to Chili. He proposed to observe there, not only the planet Venus, when near inferior conjunction (as suggested by Professor Gerling), but also the Oppositions of Mars which would occur in the years 1850 and 1852.

His proposal was accepted.

Gilles went to Chili, and there made *more than* 200 *series of observations* of Mars and Venus together. To combine with this splendid mass of work, *only* 28

corresponding observations were made in the Northern
Hemisphere, and even these do not appear to have
been made with exceptional care, nor to possess the
accuracy required in so delicate a research. The want of
sympathy and support from astronomers, which Lieut.
Gilles met with, is a blot upon the history of astronomy, and Dr. Gould, in the work I have already quoted,
has well said, " It is impossible to refrain from the
expression of deep regret that, from all the observations
of the well-equipped and richly-endowed observatories
of the Northern Hemisphere, so few materials could be
found toward rendering available, according to its
original purpose, an expedition to which so much
labour and enthusiasm had been consecrated, and to
which an accomplished observer, already known for
the precision of his measurements, had devoted his
entire energies during so long a sojourn (three years) ;
moreover, after the preparation and wide dissemination
of ephemerides and charts of the comparison stars for
both the planets during the whole period." Dr.
Gould has reduced the whole mass of observations
with a loving care, and obtains the result

$$96\tfrac{1}{10} \text{ millions of miles}$$

a result confessedly unsatisfactory from the non-agree-
ment of the various observations.

Meanwhile theorists had been at work upon the motions of the Moon and Planets,—weighing one against the other, in fact, by finding how much their mutual attractions disturb each other. Some of these disturbances or inequalities depend upon the distance of the Earth from the Sun, that is to say, if the Earth were nearer the Sun, these inequalities would be greater and *vice versâ.* It would be out of place to detail here the various results which have been derived from the application of these methods. I need only state that the results obtained about this time, by these methods, give from $91\frac{3}{10}$ to $91\frac{8}{10}$ millions of miles for the Sun's distance; instead of $95\frac{4}{10}$ obtained by Encke from the Transits of Venus.

In 1862 there occurred a very favourable Opposition of Mars. Dr. Winnecke, an astronomer of Pulkowa (the Imperial Observatory of Russia), drew up a programme of observations which was more or less perfectly carried out at six observatories in the Northern, and at three observatories in the Southern Hemisphere. Two partial discussions of some of these observations appeared, giving $91\frac{2}{10}$ and $91\frac{4}{10}$ millions of miles respectively for the Sun's distance.

Then Mr. Stone rediscussed the Transit of Venus observations of 1769, employing only the observations in which both ingress and egress were observed. He included the suspected observations of Father Hell, and by interpreting differently the language of various observers, and applying certain corrections for different phases observed, he obtained

$$91\tfrac{7}{10} \text{ millions of miles}$$

for the Sun's distance. This quantity was in satisfactory agreement with the results of the theoretical methods, and also with the recent results of the observations of Mars. Not only was this the case, but also all the observations were brought into the most beautiful accord. The largest error of observing the duration did not amount in any case to six seconds of time, and the probable error of one observation of duration taken by chance, was only two seconds of time. But the duration is made up of two contacts, so that two seconds combined the probable error of two contacts. According to the theory of probabilities, the probable error of one contact would then be *two seconds divided by the square root of two*—in other words, the probable error of one observation of contact was only one second and four-tenths of a second of time.

This result was accepted at the time with enthusiasm, and Mr. Stone received the gold medal of the Royal Astronomical Society for his labours.

About this time, however, Newcomb, the well-known

American astronomer, published his masterly discussion of the observations of Mars made at nine observatories in the year 1862. The result arrived at was

$$92\tfrac{2}{10} \text{ millions of miles}$$

for the Sun's distance. By his own rediscussion of some of the theoretical methods, Newcomb also showed that these could be reconciled with the result he found from the observations of Mars alone. He also pointed out that, according to a discussion by Powalky, the Transit of Venus of 1769 afforded a similar result, and by combining all the various methods, he arrived at the conclusion that the true mean distance of the Sun is

$$92\tfrac{4}{10} \text{ millions of miles.}$$

An additional confirmation of Newcomb's result had been previously derived by Foucault from his determination of the velocity of light.

The angular velocity of the Earth's motion round the Sun is accurately known; hence, if the Earth's linear velocity can be determined, the radius of motion (*i.e.*, the Sun's distance) will also be determined. Now, the proportion which the velocity of light bears to the velocity of the Earth's motion is pretty well determined by astronomical observation; for the fact that light takes an appreciable time to travel, has the effect of shifting the apparent places of the stars. By determining the maximum amount of this shifting (called *the constant of aberration*), the proportion of the velocity of light to the mean velocity of the Earth's motion becomes known. Thus, if the velocity of light is determined, the velocity of the Earth's motion will become known, and hence the Sun's distance.

The English astronomer, Bradley, was the first to trace out the cause of this shifting of the apparent places of the stars.

The story is that the true explanation occurred to him as he was sailing in a boat on the Thames. The wind blew directly down the river, and, when the boat was at rest, the flag also was blown in a direction straight down the river; but when he tacked to the right, the free end of the flag was carried to the left; and when he tacked to the left, the free end of his flag was carried to the right.

He had been much puzzled by changes in the apparent places of certain stars, which he could not account for. Now the truth flashed upon him suddenly.

The case is similar with my boat and its flag. The true direction of the wind represents the mean direction of a star; the boat tacking from side to side represents the Earth in its yearly revolution going from one side of the Sun to the other. The deviation of the flag from its true direction depends upon the velocity of the wind and the direction and velocity of the boat's motion. If, then, the velocity of light is not infinitely great, as compared with the velocity of the Earth's motion, the apparent place of a star must be changed according as the Earth is moving in one direction or the other; in the same way that the apparent direction of the flag is changed according to the direction of the boat's motion.

So Bradley set to work to reconcile his observations on this supposition, and succeeded perfectly, making, in so doing, the first determination of the *constant of aberration.*

This constant he found to be $20''.2$, whence he concluded that the velocity of light was 10,210 times as great as the velocity of the Earth's motion in its orbit. Succeeding astronomers have made careful determinations of the *constant of aberration,* with various results from $20''.2$ to $20''.6$.

The value of this constant which has most generally received the confidence of astronomers is the result obtained by Struve, $20''.4451$.

The Sun's distance found by Encke from the Transit of Venus, combined with Struve's value of the constant of aberration, gave 191 thousand miles per second as the velocity of light.

Later, Foucault, by most ingenious optical and me-

chanical arrangements actually determined the velocity
of light, and found it to be between 185 and 186
thousand miles per second, a velocity which, combined with Struve's value of the constant of aberration,
gave

92$\frac{3}{10}$ millions of miles

for the Sun's distance; a quantity in satisfactory
accord with Newcomb's result. It was not until 1872,
however, that Newcomb's conclusion received its most
remarkable confirmation. In that year Leverrier
communicated a paper to the Paris Academy of
Sciences, in which he gave three values of the Sun's
distance, which resulted from three independent re-
searches in the theory of the Planets.

1. From the motions of Mars   92$\frac{3}{10}$ millions of miles.
2.   „   „   Venus 92$\frac{3}{10}$ „
3.   Other motions of Venus 92$\frac{3}{10}$ „

In such estimation did Leverrier hold the accuracy of
these results, that he conceived it almost impossible
that any direct means of observation could furnish a
better determination.

He proposed, as preferable to the Transit of Venus,
the mode of determining the Sun's distance by observations for the Velocity of Light and the Constant of
Aberration, and recommended the Academy to take
steps in these directions.

Of the Transit of Venus, he says, " the determination of the Solar parallax by means of the Transit of Venus still retains all its interest, but conditionally on its being made with exceptional precision, so that the astronomer may be able to answer for it with an accuracy exceeding $\frac{1}{100}$ of a second of arc." (This accuracy corresponds with $\frac{1}{10}$ of a million of miles in the Sun's distance).

Leverrier did not believe that the Transit of Venus could yield such an accuracy; and in this respect, as well as in regard to the accuracy of his theoretical conclusions (confirmed as they were by Newcomb's results), Leverrier's opinion was shared by many astronomers.

By others, however, this latter opinion of Leverrier's was not held, but it seemed desirable to all that the Transit of Venus of 1874 should be observed in the best possible manner. The aid of photography was called in; the use of the most exquisite of all angle measuring instruments — the Heliometer — was discussed and adopted; volumes were written, and papers were read and discussed as to the best stations for observation; and all the astronomical talent of all nations busied itself in preparation.

No scientific object ever before excited such widespread activity and interest; or received from Governments and individuals such hearty and substantial assistance.

The extent of labour and of toil expended upon the Transit of 1874 may probably be gathered from a glance at the following table, prepared from the Report of the Council of the Royal Astronomical Society for the year 1876, showing the stations where the Transit was more or less successfully observed :—

Station.	Nationality.	Ingress. No. of Observations.	Egress. No. of Observations.	No. of Photographs.	Heliometers. Number employed.
Australia					
* Adelaide	British		2		
* Melbourne.	,,	1	3	200	
* Sydney .	,,	2	2	180	
* Windsor .	,,	1	1		
Burmah					
Bamo .	?	1	1		
Beyrout .	?		1		
Cape of Good Hope	British		3	14	
Ceylon .	,,				
Columbo .	,,	1	1		
Egypt					
* Cairo	,,		3		
Gondokoro .	,,		1		
* Suez.	,,		1		
* Thebes .	Russian		1		
* Thebes	German		1		1
* Thebes .	British		1	35	
India					
Bushire.	,,	1	1		
Calcutta .	,,	1	1		
* Kurrachee	,,				
* Maddapore	Italian	1	1		
* Roorkee	British	2	2	100	

Station.	Nationality.	Ingress. No. of Observations.	Egress. No. of Observations.	No. of Photographs.	Heliometers. Number employed.
Islands					
* Auckland . . .	German	I	E		1
* Bourbon . .	Dutch		3		1
* Chatham . . .	American			8	
* Sandwich Islands .	British	5		60	
* Kerguelen . .	,,	2	2		
* Kerguelen . .	German	I	E		
* Kerguelen . .	American			26	
* Mauritius . .	German		2		1
* Mauritius . . .	Lord Lindsay's Exp.		3	100	1
* Mauritius . .	British	1	1		
* New Caledonia. .	French	1		100	
* Rodriguez . .	British	3	3		
* St. Paul . .	French	I	E	500	
Japan					
* Kobe . . .	,,	I	E	P	
* Nagasaki . . .	American	I	E	60	
* Nagasaki . .	French	I	E	P	
* Yokohama . .	Russian	I	E		
* Yokohama . .	Mexican	2	2	100	
New Zealand					
* Bluff Harbour . .	German	I		P	
* Christchurch. .	British			9	
* Queenstown . .	American	I		59	
Persia					
* Ispahan . .	German			19	
* Teheran . . .	Russian	I	E	P	
Russia and China					
Habarowka . .	,,	I			
Jalta . . .	,,		E		
* Kiachta . .	,,	I	E	8	
* Nertschinsk . .	,	I	E		1
Orianda . .	,,		E		

Station.	Nationality.	Ingress. No. of Observations.	Egress. No. of Observations.	No. of Photographs.	Heliometers. Number employed.
Russia and China—*cont.*					
* Port Passuet . .	Russian	I	E	38	
* Pekin . . .	American	I	E	90	
* Pekin . . .	French	I	E	60	
* Cheefoo . .	German	I	E	P	1
* Tschita . . .	Russian	I	E		1
* Wladiwostok.	,,		E		
* Wladiwostok . .	American	I	E	13	
Tasmania					
* Campbell Town .	American		E	55	
* Hobart Town . .	,,			39	

In all cases where I am not sure how many observations were obtained, I have substituted the letters I and E to represent an unknown number of observers of Ingress or Egress respectively. Also, where the number of Photographs obtained is unknown, I have substituted the letter P.

With regard to the photographs, I have simply stated the number of pictures reported as obtained. It is only when these have been tested under the microscope that it is possible to say how many of them will really be useful for measurement—probably not more than one-tenth of the number will prove to be so.

At all the stations marked with an asterisk, some at least of the observers were men with previous training,

provided with thoroughly good instruments; indeed it
is very questionable if any observations made without
these conditions should be admitted into the final
discussion.

Now what are the results of all these observations?
It is as yet impossible to say with anything like certainty, but, from partial discussions which have been
published, we may draw some conclusions.

The photographic observations have resulted in
failure, at least so far as the British stations are
concerned. The pictures have not the sharpness
necessary for the delicate measurement to which they
must be subjected, and they appear besides to be
affected by systematic errors inherent to the method
employed.

In criticising this result, it must not be forgotten
that the method is experimental; that its first application to the measurement of small angles (requiring
an accuracy of $\frac{2}{100}$ or $\frac{3}{100}$ of a second of arc) was on
the occasion of the Transit of Venus, and that no satisfactory means existed of putting the method to
previous proof.

With the experience of last Transit, however, I
think it is not impossible that photography may
be successfully employed to observe the Transit of
1882.

The Heliometer observations are not as yet published in a form from which any accurate estimate of their value can be formed.

The eye observations of contact present the same difficulties as in former Transits.

The contact is not a sharply-marked phenomenon, but a gradual merging of two limbs. It is complicated by the effects of the atmosphere of Venus, by the irradiation of the Sun's limb, and by the undulations of our own atmosphere. These effects also vary with the description of telescope and accessories employed; with the depth and colour of sun-shade; with the aperture of the telescope and its magnifying power; and, lastly, with the imagination of the observer, with his previous impression as to what he ought to see, or hopes to see, and with the language in which he describes what he believes he has seen.

Picture for a moment the circumstances under which the observations are taken. The observer has made a long, it may be a perilous journey, and has perhaps encountered much difficulty in landing and erecting his instruments and Observatory. The weather for days before the 9th of December has been unpropitious, and the observer is worn out by long and anxious watching. But when the eventful morning arrives, the sky is cloudless, and the Sun shines in all his strength. The

astronomer is exultant, and the revulsion of feeling sets
his every nerve a-tingling.

Soon after the predicted time, he sees a small black
indentation in the Sun's limb. The indentation increases. The black disc of the Planet has half entered
upon the Sun—when, what is this? Around a portion
of the black disc appears a band of light, that extends
till it forms a ring round the Planet. Three minutes
more and the critical instant will be past; and yet the
mystery of the unexpected ring remains unsolved.
Meanwhile the definition becomes less and less satisfactory, for the strong Sun has heated up the rocks
which surround the Observatory, and tremulous currents
of air ascend which render the image blurred and ill-
defined. About this time, according to the observer's
previous experience with " the model," fine sharp cusps
should be rapidly approaching each other, and, when
they meet, light should appear between the edge of the
Planet and the Sun. But no! the cusps are not fine;
they are blunted and rounded off; and this bright ring
of light complicates the matter.

A few seconds more, and all will be over. Even yet
the observer does not know what he ought to note.
He feels that *now* is the supreme moment, that *now*
he must reap the fruit of all his labour, or lose his
harvest.

Still there is nothing precise to observe. So, in the tremulous image before him, he notes what best he can,—and then the Transit is over.*

After the first feeling of disappointment, comes a certain feeling of satisfaction. The important contact has been observed. The circumstances were favourable. All has been done that could have been done. Thus the report unconsciously partakes of a more buoyant character than would have been the case could the observer precisely recall his feelings at the instant of observation ; or than if he knew that to-morrow, and

---

* I do not think I have exaggerated the difficulty and uncertainty of the observation in the slightest degree, and I quote in corroboration the words of a most conscientious observer :—

" Sky perfectly clear ; no cloud.

" On focussing, after changing the micrometer, to my astonishment I saw a completion of light round the planet, perfectly distinct, and such as I should have said, from previous model-practice, was immediately after contact. This is the time recorded. I remained looking at it for about two minutes, but could see no instantaneous phenomenon of contact, no black drop, nor anything resembling the model. I noticed that this light did not appear to thicken as I should have expected for a considerable time after that recorded, but as I considered, from my previous experience [with the model], that the contact had occurred, and was unable to get, accurately, any further change until the planet was visibly on the Sun, I cannot say that the time as noted is at all satisfactory."

Had all observers been equally hard to please as to the precision of the phenomenon, the result of the Transit would have been—no observations. But these words will undoubtedly recall to many an observer the unsatisfactory character of the phenomenon which he had to note.

the next day, and the next, he could repeat his observation again and again.

Is it then a wonder that any person who attempts to select " corresponding phases " of the Transit, from the uncertain and incongruous records of the various observers, should vacillate in his opinion as to the true interpretation of their language ? Is it a wonder that anyone who has previously formed a strong opinion as to what the result should be, and who has made himself familiar with the effects of different interpretations on the final figures, should unconsciously give such interpretation as will lead to a result agreeing with his preconceived ideas ?

I do not think that it is ; the wonder would rather be if it were otherwise.

Accordingly we find that various results have already appeared. The first of these was obtained by M. Puiseux from the French observations at Pekin and St. Paul : it gave

$92\frac{3}{10}$ millions of miles

for the Sun's distance, and so far confirmed Newcomb's and Leverrier's conclusions. The next combination, however, gave a very different result—more nearly Encke's distance. So the French, like the Germans, very wisely resolved to publish only the observations made, without drawing any conclusions from them,

leaving the definitive result to be deduced from the combined observations of astronomers of all nations, according to the recommendation of our Astronomer Royal.

This was indeed a wise recommendation, for partial
discussions can tend only to hinder true progress.

The British public, however, as is their manner,
upset this wise resolution, and Parliament demanded
immediate value for its money. " What is the result?"
was the impatient cry of honourable members, doubtless waiting with restless anxiety for information as to
the Solar Parallax. And so the Astronomer Royal
had to prepare a Blue Book, and give the result of the
British expeditions alone. The result contained in
this report was

$$93\tfrac{4}{10} \text{ millions of miles}$$

for the Sun's distance.

Then Mr. Stone (H.M. Astronomer at the Cape)
rediscussed the same observations, and deduced

$$91\tfrac{8}{10} \text{ millions of miles}$$

or a result differing by more than one and a half
millions of miles from the Astronomer Royal's result.

Subsequently, introducing a wider range of observations, the Astronomer Royal announced in his Report
to the Board of Visitors, that the observations seemed
now to point to about

$$92\tfrac{7}{10} \text{ millions of miles.}$$

Finally Captain Tupman, the chief of the British Expeditions, from a still more extended discussion deduced

$$92\tfrac{5}{10} \text{ millions of miles,}$$

a result agreeing precisely with that of Newcomb and Leverrier.

With admirable candour and fairness, however, he states

" Although the above results " (referring to the results from Ingress and Egress taken separately), "present such an unexpected agreement, it cannot be said that the *mean*" (equivalent to a Sun's distance of

$$92\tfrac{4}{10} \text{ millions of miles)}$$

" is entitled to much confidence."

      *   *   *   *   *   *

" Any selection of times *made after the investigation of the effects of parallax*, will always expose the result to the suspicion of having been ' doctored.' "

In these opinions I heartily concur.

Of course, no determination of the Sun's distance is likely to be correct which is irreconcileable with a reasonable interpretation of a well-observed Transit of Venus—like that of 1874 ; and therefore, as a *confirmation* of *other results*, the eye observations of contact may prove useful. But of this I am convinced, that no eye observations of a Transit of Venus can ever satisfactorily *determine* the Sun's distance; though it is not impossible that photography, properly employed, may accomplish the desired end.

One very useful lesson taught by this discussion of the Transit of 1874 is, that the apparent agreement of the observations, in Mr. Stone's discussion of the 1769 Transit, is entirely illusory. In the Transit of 1874, the probable error of an observation of contact amounted, at least, to 7 or 8 seconds of time. These observations were made with the best modern appliances, very, very far superior to any of those used in 1769. From what we know of the instruments, and the nature of the observation, the real probable error of contact in 1769 must have been at least 10 seconds (instead of one and a half), and the resulting distance of the Sun may be anything from $90\frac{1}{2}$ to 95 millions of miles.

But between the observation and discussion of the Transit of Venus of 1874, results of other interesting investigations were published.

In 1872 Dr. Galle of Breslau had proposed a series of observations on the minor Planet *Flora*, at its Opposition in the autumn of 1873, for the purpose of determining the Sun's distance. He contended, that, though this Planet would not approach so near to the Earth as Mars and Venus do in certain circumstances, yet its minute disc, exactly like a Star, would form a better object for exact measurement, and one less liable to systematic error of bisection. He secured the co-operation of Observatories in the Northern and Southern Hemispheres, and, by combining the observations so obtained, he derived

92½ millions of miles

for the Sun's distance.

This result was also confirmatory of the Newcomb-Leverrier value, and the method offered great probability of freedom from systematic error. But at some of the most important Observatories the instrumental equipment was not satisfactory for so delicate an inquiry; and the result therefore hardly possesses the importance which it would otherwise have.

Meanwhile the French Academy had adopted the suggestions of Leverrier, and M. Cornu was selected to redetermine the Velocity of Light. The investigation was executed with eminent skill and care, and the result, combined with Bradley's determination of the constant of aberration, gave $92\frac{2}{10}$ millions of miles for the Sun's distance, a result also in agreement with that of Newcomb and Leverrier; but when combined with the far more refined and more generally accepted determination of Struve, the result is

93 millions of miles

Such was the state of our knowledge of the Sun's distance in the beginning of 1877.

Opinions were divided. Few, if any, still adhered to the old value of Encke ($95\frac{4}{10}$ millions of miles), but some firmly maintained the accuracy of the other extreme ($91\frac{6}{10}$ millions of miles).

The vast majority of Astronomers adopted the Newcomb-Leverrier value ($92\frac{2}{10}$ millions of miles), but very few believed the constant to be yet definitively established. It was still possible that these two coincident results might each be subject to a small systematic error; and some investigation, to which systematic error could not possibly be attributed, was earnestly desired.

In 1857 the Astronomer Royal had proposed the mode of observation described in Chap. I. p. 9, and gave it as his opinion that it was the best of all methods to determine the Sun's distance.

This proposal had never been satisfactorily carried out, and yet it offered many advantages. It required no co-operation, and the whole of the observations might be made by the same observer with the same instrument, so that systematic errors would be entirely avoided.

Combining the suggestions of the Astronomer Royal and of Dr. Galle, Lord Lindsay and I proposed in

1874 the observation of a Minor Planet in the evening and early morning, as the best method of determining the parallax; and we showed that, by employing the Heliometer in the observations, there was a probability of realizing a higher accuracy than had ever before been attained. The practical form which the proposal took, was to observe the Planet Juno on the occasion of Lord Lindsay's Expedition to Mauritius; and this was duly done. On account of the late arrival of Lord Lindsay's yacht at Mauritius, only a very small proportion of the intended observations were secured; but these, on reduction, proved to demonstration the extreme accuracy of the method. Though the observations are not numerous enough to give a determination of the Sun's distance of the highest precision, it is still interesting to find that the tendency of the result was to confirm the Cornu-Struve result derived from the Velocity of Light and the Constant of Aberration.

From all the observations of Juno combined, the result was

$93\frac{3}{10}$ millions of miles,

or rejecting one outstanding result

$92\frac{8}{10}$ millions of miles.

The Opposition of Mars in 1877 offered, so far as geometrical conditions are concerned, the most favourable opportunity of the century to determine the parallax, by observations at a single station.

I thought it would be a matter of the greatest regret if such an opportunity were lost. Having mentioned the matter to Lord Lindsay, he, in the kindest manner and in fullest sympathy with the importance of the object, at once placed his Heliometer at my disposal.

I had already much experience in the use of this instrument, and had spent many months in finding out its errors and the best means of correcting them. This great labour therefore would not have to be repeated.

The combination of circumstances was altogether so fortunate, that the Royal Astronomical Society, on the earnest recommendation of the Astronomer Royal, guaranteed the £500 which I considered necessary for the expenses of the expedition. Through the good offices of the Astronomer Royal, I also received such letters from the Lords of the Admiralty, and such efficient assistance was in consequence given to me at Ascension, that the money voted by the Society proved more than sufficient. Numerous evening and morning observations of Mars were secured; and the reductions, now far advanced, promise a result of very great accuracy.

I may state here that the observations of one week (Sept. 4—11), which are now reduced, confirm the tendency of the Juno result and of the Cornu-Struve value of the Sun's distance ; but a good many months must still elapse before the final result, from all the observations, can be deduced.

Another Astronomer, Mr. Maxwell Hall, of Jamaica, observed Mars exactly on the plan of the Astronomer Royal. The details of his observations are not yet published, but the result he arrived at also confirms the tendency of the Cornu-Struve, Juno, and Mars results.

Without entering into greater detail, I may state that, if these recent results are confirmed, the Sun's distance will prove to be nearer to 93 than to 92 millions of miles.

It may appear strange to the uninitiated that Astronomers should be in doubt about so large a quantity as a million of miles, but perhaps a familiar illustration will convey some idea of the difficulty of the problem.

The apparent size of the Earth, looked at from the Sun, is about that of a globe 5½ inches in diameter viewed at a mile distant.

If this 5½ inch globe is shifted 57 feet nearer to the observer, it will be increased in apparent diameter just as much as the Earth would be if shifted a million of miles nearer to the Sun—or, as if the $5\frac{1}{2}$ inch globe had not been shifted, but had been increased by $\frac{6}{100}$ of an inch in diameter. Hence, in measuring the parallax, an angular error, corresponding with $\frac{6}{100}$ of an inch viewed at a mile distance, will produce the error in question.

If any one desires to form an adequate idea of the difficulties of measuring the Sun's distance to a million of miles, *let him try to measure the thickness of a florin-piece, looked at, edge on, a mile off.*

I have endeavoured in the preceding sketch to outline the History of the great Problem which occupied our time and thoughts at Ascension. I shall be well satisfied if I have enabled any one to realize somewhat of its nobility and interest, and the consequent intensity of our anxiety for a successful result.

DAVID GILL.

LONDON: *November*, 1878.

CONTENTS.

CHAPTER III.

ST. HELENA.

CHAPTER IV.

WHAT ASCENSION LOOKED LIKE.

CHAPTER V.

ASCENSION PAST AND PRESENT.

CHAPTER VI.

ROUND ABOUT GARRISON.

CHAPTER VII.

A NIGHT ON THE CLINKER.

CHAPTER VIII.

CHANGE AND CHECK.

CHAPTER IX.

MARS BAY.

CHAPTER X.

A SUNDAY SCENE.

CHAPTER XI.

THE OPPOSITION OF MARS.

CHAPTER XII.

THE SEA-SHORE AND THE ROLLERS.

CHAPTER XIII.

GREEN MOUNTAIN.

CHAPTER XIV.

SUNDAY AT THE MOUNTAIN.

CHAPTER XV.

WHY WE HAD ONLY A GALLON OF WATER.

CHAPTER XVI.

TRIPS FROM GARDEN COTTAGE.

CHAPTER XVII.

MARS BAY WITHOUT A COOK.

CHAPTER XVIII.

WIDE-AWAKE FAIR.

CHAPTER XIX.

LAST DAYS AT MARS BAY.

CHAPTER XX.

CHRISTMAS IN GARRISON.

CHAPTER XXI.

ABOUT THE KROOMEN.

CHAPTER XXII.

CLINKER CEMETERIES.

CHAPTER XXIII.

CHRISTMAS HOLIDAYS.

CHAPTER XXIV.

THE DEVIL'S RIDING SCHOOL.

CHAPTER XXV.

HOMEWARD BOUND.

SIX MONTHS IN ASCENSION.

CHAPTER I.

WHY WE WENT.

I REMEMBER a story once told me by a learned friend. He had been explaining to a lady, with much care and minuteness, the reasons why the axis of the earth is slowly though constantly changing its direction in the heavens, and why, therefore, the star, which is the Pole star now, was not the Pole star 4000 years ago.

The lady had encouraged our friend to proceed with his explanation by the most marked attention, and by such appreciative interjections as "Really!" "Indeed!"

B

"How beautiful!" In this way he was led to more than usually minute description, and with much unction proceeded to crown his argument as follows.

"Now you see, by this change of the direction of the earth's axis, if we have any permanent record of an observation of the angular distance of a star from the Pole, we can calculate how long ago that record was made." "Of course!" "And in the Great Pyramid we have such a record." "Indeed! how wonderful!" "The entrance passage points to the north, and its angle of inclination corresponds with the lower culmination of the Pole star of 4000 years ago."

Here a little hand was laid on our friend's arm, and his feelings may be better imagined than described, when, in an anxious voice, the question was put, "And pray, Professor, *what is an angle?*"

Now, I too have a story to tell in which angles occur, and, warned by the Professor's experience, I would leave them out if I could. This, however, I cannot do altogether, lest some should thus miss the point of the story; but, as next best, I shall throw them all into the first chapter; and those of my sisters who care for none of these things, or who, like the Professor's fair friend, know not the meaning of an angle, may pass it over and read about our "Six Months in Ascension," without the reasons that took us there.

It was no longing for new scenes, no thirst after gold, no need for better health that led us to bid

good-bye to England in the leafy month of June, and
seek a barren rock at the Equator. It was none of
these things. We went in search of the Solar Parallax, or in other words to find out how far off the sun
is from the earth.

Many a noble head has puzzled over this problem,
and many a sage thought and many an hour of careful
observation did the grand old philosophers give to its
solution. Our own Astronomer Royal, in a paper read
in 1857, says : " The measure of the sun's distance has
always been considered the noblest problem in astronomy." The general interest taken in the last Transit
of Venus, and the large sums expended by different
nations in providing for its observation (amounting in
the aggregate to about a quarter of a million sterling)
show that the solution of the question still maintains
its importance, not only as the settlement of an ab-
stract truth, but as an essential condition to the future
progress of astronomy; and I think I can show why
this is so.

The astronomer knows, and has known for ages,
the proportional distances of all the bodies of our
solar system; he knows too, as a mathematical fact,
that there is an exact relation between the time which
a planet takes to make a complete revolution round
the sun and the distance of that planet from the sun.
Now it is easy to find the time which a planet takes to
go round the sun, and the knowledge of the old astronomers in this respect was nearly as accurate as that
of our own day.

When Kepler discovered his famous law of the distances,* he was in a position to draw nearly as accurate a chart of the sun and the paths of the heavenly bodies as we could draw at the present time. What he could *not* do was to give the scale of his chart. He could say at any time, "If you can tell me what is the distance between any two planets I can tell you the distances of all the others, because I know exactly the relative proportions of all these distances."

He was, in fact, in the position of a man who has a map of England placed before him and is requested to tell from it the distance from London to York. If the map has no scale attached to it he cannot do so : but if he is told that one inch on the map is equivalent to ten miles, he has only to measure the number of inches between London and York on the map and to multiply the result by ten, to find the distance in miles.

Or again, if he happens to know any other distance on the map, such as the distance from London to Oxford, he has only to find how many times this distance is contained in the distance from London to York by the map, and thus he solves the question.

Accordingly, the astronomer requires only to know one distance in our solar system in order to know all; and for this purpose he selects as his unit of measure the mean or average distance of the earth from the

---

* The squares of the times of the revolutions of the planets are proportional to the cubes of their mean distances from the sun.

sun. Thus we do not order 9000 millimetres of
silk for a dress, nor astonish the draper by demanding
$\frac{1}{176}$ of a mile ; but prefer to ask for 10 yards of
silk. So the astronomer, instead of saying that the
average distance of Jupiter from the sun is 478
millions of miles, prefers to say its distance is $5\frac{1}{5}$,
meaning that it is $5\frac{1}{5}$ times the distance of the earth
from the sun.

The determination, then, of the length of the astronomer's unit, is of the same importance to him, as is
the true length of the yard measure in the ordinary
business of life, or in the more scientific work of the
surveyor or engineer. In order to accomplish this
determination, he has, as I have shown, only to find
the distance of any one planet from another ; and
now I must explain how this is done.

Many will remember the great meteor-shower of
1866. On that occasion, one very remarkable meteor
was seen by an observer in Aberdeen to burst in the
South, apparently near a well-known star in the con-
stellation of the Bull, while to another observer in
Newcastle the same meteor appeared to burst in the
North, near another well-known star in the Great
Bear. The time of bursting, and the angular distances of the stars (from the Pole and from the
meridian) being known, it was easy for the astronomer
to calculate the apparent altitude and direction of
these stars as seen from Aberdeen and Newcastle
respectively.

But without the employment of mathematical terms,

it is difficult to explain how he works; and with them, it would be hopeless for me to attempt the explanation. I myself do not understand mathematical terms, so how could I use them with the hope of explaining these things to my readers? However, I can use knitting-needles, and perhaps they may do just as well.

Let us suppose that the astronomer takes a map of England and places one end of a knitting-needle on the town of Aberdeen, then turning the other end of the needle in the proper direction he raises it to the required altitude for the star in the Bull. Similarly, he takes another needle, places one end on Newcastle, turns the other end in the direction of the star in the Great Bear, raises it to the required altitude, and where the needles cross each other, *there* must be the place of the meteor. In the case in question it was found to be 40 miles vertically over the town of Dundee.

Of course the astronomer uses no such clumsy contrivances as knitting-needles. He finds the lines of the mathematician more convenient, but the principle is the same. To measure the distance of any celestial body from the earth, it is only necessary to observe it from two different points. In the case however of measuring the great distance of a planet, the problem becomes very difficult. For even when the planet is looked at from opposite sides of the earth, the lines (or needles) must go so far before they meet, that the angle at the apex is almost insensibly small, and yet on the measurement of this minute angle (called the parallax*) does the whole problem depend.

The Transit of Venus had been supposed to afford the most accurate means of measuring a planet's distance; because at a Transit of Venus the planet is only at about one-fourth of the sun's distance from the earth, and passes across the sun between it and the earth. At that time, when the edge of the planet seems to an observer on one side of the earth to touch the edge of the sun, to an observer on the opposite side of the earth this does not appear to be the case, because of the apparent change in the planet's position, produced by its being viewed from different points.

In this way, to observers situated at different points of the earth, the edge of Venus appears to touch the edge of the sun at different times, and from the difference of these times astronomers can calculate the angular change in the planet's apparent position. Then, since they know the size of the earth and the latitudes and longitudes of the observers' stations (and consequently their distance apart), they are able to calculate the distance of Venus in the same way that the distance of the meteor was calculated.

But a somewhat unexpected difficulty was found in the observations of the Transit of Venus. It was discovered that the planet Venus was surrounded by a

* Astronomers in practice define the parallax as half this angle — that is, the angular amount that the object is displaced from its position as seen from the centre of the earth.

dense atmosphere, so that the sun's edge seen through it was hazy and indistinct. The observations consequently were not made with the precision that was necessary; and it became desirable to find some other method of settling the great problem.

Now, it happened that during August and September of 1877, the most favourable Opposition of Mars possible in the present century would occur. An " Opposition of Mars" occurs when that planet, the earth and the sun, are nearly in a straight line, the earth being between the planet and the sun. Hence the planet comes to the meridian at midnight. In the case of Mars, this condition of things is realized nearly every two years, but at the Opposition of 1877 he would be nearer to the earth than at any Opposition during the present century, and on the 5th September he would be only one-third of the sun's distance from the earth.

Under these circumstances my husband proposed a method of observing the planet which he believed to possess advantages over all other methods. Instead of employing two sets of observers in different parts of the earth, as in the Transit of Venus, he determined to combine both sets of observers in himself. He would thus avoid the disappointment that occurred to Lieut. Gilles, who went to Chili in 1850 and made laborious observations for the parallax of Mars, but found on his return that hardly any corresponding observations of importance had been made in the northern hemisphere. Thus his labour had been

expended in vain. The way in which my husband
managed to avoid the possibility of a like catastrophe
was as follows.

He proposed to observe the planet in the evening
when it was rising—in other words, to look at it
from position A; then to observe it in the early
morning when it was setting, that is to
say, to observe it from position B. He
availed himself in fact of the rotation
of the earth to carry both himself and
his Observatory round, and so, by merely
waiting, to be transported 6,000 or 7,000
miles between the times of his evening
and morning observations.

This part of the proposal was not
original. It had been suggested long
ago by the Astronomer Royal, but had
never been acted on. My husband, however, proposed for the first time the *method* of observation
which he carried out, *viz.*, the employment of the
Heliometer—the most exact of all angle-measuring
instruments—and, to secure accuracy in the result, he
elaborated details which need scarcely be described
here.

When his scheme was ripe, he drew it up in complete
form, and it received from the Royal Astronomical
Society, from the Astronomer Royal and others, the
most cordial support. A sum of 500*l.* was granted
from the funds of the Royal Astronomical Society in
April, 1877, in order that he might carry out an

expedition to the Island of Ascension, to observe the Opposition of Mars in the following autumn.

This expedition, however, no *money* could have rendered possible at this date, had not Lord Lindsay, with the greatest kindness, agreed to lend for the purpose his splendid Heliometer. This is the only instrument of the kind in England except the much larger one in the Radcliffe Observatory, Oxford, and it was not available.

My husband decided on Ascension as the most suitable station for making the desired observations, on account of its favourable position with regard to latitude, and its reputed meteorological conditions. Through the good offices of the Astronomer Royal, the consent of the Admiralty was obtained for our occupying this station.

The Transit Instrument was lent by the Royal Astronomical Society. Some considerable modifications were necessary; but these were duly made. The Astronomer Royal lent the Transit Hut; the Heliometer Observatory, a chronograph, five chronometers, two reflecting circles, some barometers and thermometers, completed our instrumental equipment. The whole, in their packing cases, together with our personal luggage, made up about 20 tons measurement of baggage.

But before starting, very particular attention was required in regard to the Heliometer—the keystone on which the whole structure of the work rested. And here begins the story of its adventures and mischances.

The instrument had never been used in so low a
latitude as Ascension, and it was necessary to test it
carefully, in order to ascertain whether it would perform its functions well under the untried circumstances.

Considerable interest in the expedition having been
shown by members of the Royal Astronomical Society,
it was thought best to erect the instrument in their
rooms at Burlington House, where the necessary trials
could be made, and that the instrument might afterwards be exhibited and explained at one of the evening
meetings. The Heliometer was duly erected and all had
been brought nearly into the same condition of affairs
as would be required at Ascension. David was applying a level to an inclined piece of wood cut to the
angle of the latitude of Ascension, and was directing
the workmen to give a final motion to the screw by
which the inclination of the axis is changed, when
slip! the screw gave out, the overhanging weight of
the Heliometer and its counterpoises tore the lower
end of the cradle from his hand, and, tilting upwards,
the polar axis, counterpoise weights and Heliometer-
tube, in all several cwt., came down crash, from a height
of 7 or 8 feet, upon the floor.

Imagine the astronomer's feelings as he saw the
Heliometer of all his hopes light upon its delicate eye-
end; that eye-end driven through the floor and slowly
torn off, as the whole mass gradually turned round,
smashing and crushing the more delicate rods, handles
and other attachments to the tube, and finally squash-

ing one of the copper caps which protect the ends of the slides from dust.

As the whole thing lay there on the floor, within ten days of the time when it must be packed for shipment, it seemed impossible that it could be restored fit for use. The apparent ruin of so many hopes and plans was paralysing, and for some minutes David was quite incapable of examining the amount of damage done. By-and-by, however, as he came to look into details, matters did not prove to be so desperate as they had at first sight appeared. The tearing and smashing and crushing of the eye-end, handles, &c., had had the happy effect of breaking the fall; and on removing the head, he was delighted to find that the object-glass, the slide, the scales, and in fact all the really vital parts of the Heliometer proper were intact, and working as smoothly and beautifully as ever.

The life was still there, and the shattered limbs were at once placed under the care of able surgeons, who in six days made them whole as before. Messrs. T. Cooke & Sons, the great opticians of York, Mr. Browning, and Messrs. Troughton and Simms of London, were all pressed into the work, and with a will they accomplished it. But what a time of strain it was, and how tired we were before we started! Yet all the while we never ceased to congratulate ourselves on the misfortune having taken place when and where it did.

The cause of it was simply that the elevating screw was too short, and the instrument being called a

" Universal Equatorial," that is, adapted to all latitudes, this deficiency could not have been anticipated.
Had it not been for this trial in Burlington House,
in all probability, a like accident would have happened at Ascension, the result of which would simply
have meant the utter failure of the expedition.

It was only at the last moment that we were ready;
but we *were* ready. The evil that is past is not to
come.

CHAPTER II.

THE VOYAGE.

Dartmouth.—Dartmouth shopkeepers.—Dr. Davidson's "Perfect Remedy for Sea-sickness."—Madeira.—Diving boys at Funchal.—Lordly colonists.—Teneriffe.—What the stewardess thought of the Peak.—Our fellow-passengers.—South African Mission Work.—Eaves-dropping.—Arrival at St. Helena.—The Ladder.—Landing.—"Derelicts."—Johnson's Observatory.—Its present uses.—The St. Helena Astronomer.—St. Helena skies.—Tempted to stay.

At Dartmouth, on the 14th of June, we joined the *Balmoral Castle,* a beautiful new steamer of the Donald Currie Line, bound for the Cape of Good Hope. She had left the London Docks three days before, having all our goods on board except the chronometers, which we brought with us. None of the outward-bound English mail ships touch at Ascension, so that we were under the necessity of taking our passage to St. Helena, there to wait for a return steamer from the Cape to take us back to Ascension.

After a rapid railway journey, we enjoyed a quiet night in harbour. Next morning we had a stroll through the picturesque little town of Dartmouth; for our ship would not sail until mid-day on the 15th. What a curious old town it is!—with its steep, narrow streets, shaded and cooled by projecting piazzas and gable-ends. Bewitched by the quaintness of the place, one sees the streets again alive with the brave army of Red Cross Knights that sailed from this fair bay so many hundred years ago to do battle against the Saracen, and pictures many a "ladye bright" peeping shyly from the little latticed windows to wave a last adieu to her own true knight.

Then the scene changes, and gay groups of handsome cavaliers lounge in doorways and at street corners, whilst fitful snatches of light song bring dark frowns to the grave faces of the Puritan townsfolk, as the "Merrie Monarch" holds gay court in that uncourtly-looking little mansion in the Butterwalk, where the arms of the second Charles are still to be seen, carved in oak over the fireplace.

Dartmouth is delightfully old-fashioned. It is a little romance in stone, and primitive to a degree. We thought that we had quite finished our shopping in London; but, as always happens, some odds and ends had been forgotten, and we now tried to supply them here, though not very successfully. One tiny shady shop we entered in search of a deck chair, but could find no one to attend to us for ever so long, notwithstanding vigorous stamping and knocking; until at last a neighbour pointed out that I must pull a bell-rope over the door.

This by-and-by fetched the laggard shopkeeper, who seemed to fall into a drowsy state again, while we poked about in his little shop; and, since he made not the slightest attempt to recommend any of his goods, I

suppose we cannot blame him that the chair we finally
hit upon came to pieces before we reached Madeira.

Again on board, we found everywhere the signs of
departure, and I enjoyed watching the bustle. Heavy
boxes were being lowered into the hold by strong
arms, and now and again Jack stops to refresh himself with a grim joke. "Heave again, Bill! I guess
some old fellow must be taking out his gravestone in
that 'ere box, it is so precious heavy. Try again.
Now easy. Lower away!"

At last all was stowed away, the hatches were closed,
the gangway raised, and, punctual as a mail train, we
were off at noon.

My husband had been fortunate enough to get an
empty cabin on deck for his chronometers, and when
the pilot had left the ship and England could be seen
no longer, he retired to compare, and I to wind them
—a duty I had undertaken during the voyage.
But somehow I couldn't find the keyhole ; and had
the winding of the chronometers been left to me,
I fear that they would all have run down before we
reached Madeira. Dr. Davidson's "Perfect Remedy
for Sea-sickness," with which I had carefully provided
myself, was of no use at all ; and the first three days
of our voyage presented to me little variety and less
pleasure.

On the evening of the fourth day we sighted
Madeira, and the much-worked screw had an hour's
rest, not to speak of the rest to some others, who had
been less usefully employed, but were sadly more

wearied than this steam-driven giant. Approaching
Madeira from the north, she greeted us with a stern
inhospitable face, rugged and determined in expression, with a baked, sun-burnt complexion. But as our
ship glided slowly along, the coast gradually betrayed
a gentler nature, until, turning a sharp headland suddenly, Funchal lay before us in beautiful panorama,
filling the bay and stretching upwards to the very top
of the steep background in little offshoots of white
verandahed villas and green terraced gardens.

Much to my disappointment we had no time to land :
but, immediately we came into harbour, numbers of
little boats put off from the shore. The first contained half-a-dozen of the well-known amphibious
diving-boys, vociferating in all the English their Portuguese tongues would admit of, " Haiv sare ! " "Fare-
away sare ! " " Dive for a shillin' sare ! " and down
they plunge, invariably bringing up, either in hand,
mouth or toes, the tiniest silver piece that is thrown
to them from the ship's side. Copper is scorned by
these clever little rascals, and it was amusing to wit-
ness their rage when they found that breath and richer
harvests had been lost in bringing up a penny. Had
Portuguese not been to us an unknown tongue, I fear
we should have heard some hard names.

Other boats soon crowded alongside, and, after the
necessary formalities with the Health Officer, we were
boarded by a motley crew of gipsy-looking men eager
to sell us photographs, shawls, feather-flowers and
fruit. " Sell you a basket of cherries for six shillin'

sare!" "Beautiful photograph—only three shillin'!"
and some reckless, heavy-pursed colonist, whose box
of presents from England is not quite full, lets himself be robbed in the most lordly fashion by these
barnacles, while more patient passengers get plenty of
cherries for one shilling when the last bell is ringing.

By the light of a clear soft moon we steamed off
again, and I, for one, was sorry to leave this vine-
clad isle behind me, and heard with regret the voices
of its noisy traffickers melt away in the distance.

Towards sunset on the 20th we sailed through
the Canary Islands, passing within twenty miles of
Teneriffe. As I sat on deck during dinner I feasted
my eyes on the Great Peak. At first it was almost
entirely enveloped in mist, which gradually cleared
away from the top and the bottom, till at last a single
vapoury band cut the mountain in two, producing a
wild weird effect. The dark sharp-pointed Peak,
looking as if it had nothing to rest upon, seemed to
float in the air like a lost world, and from its great
height it appeared so close as to be in danger of falling
upon us! Gradually the encircling band narrowed,
then broke up, and the little clouds floated hither and
thither, giving the most charming effects of light and
shade as they played on the rugged sides of the great
rock. Slowly these lakelets of vapour grew less and
less, then altogether vanished, and the Peak stood
revealed in its naked grandeur.

It was a splendid transformation scene, and watching it I had forgotten my dinner, till the homely old

Scotch stewardess interrupted my reverie with a plate of currant tart. I made some remark to her about the beauty of the scene. "Ay, ay," she said, "it's a big hill, but there's nae scenery in earth or ocean like oor ain Scotland."

However, putting aside the claims of our own crags, Teneriffe is a grateful sight to the water-wearied eye; and it had another interest to us besides that which its natural beauty and grandeur excited. It was here that Professor Piazzi Smyth lived above the clouds and wrote his charming book, "An Astronomer's Experiment." The story had interested me very much when I read it some years ago, the more so that the author was a valued friend, and I had now great pleasure in speculating where Guajara might be, and where the path that the astronomer and his wife had toiled up with their heavy instruments to "Alta Vista," the site of their home and Observatory for the time, 11,000 feet above the sea.

Leaving the Canary Isles behind us we saw no more land, with the exception of a distant glimpse of Cape Verde, until we reached St. Helena. But to compensate for this, the sociability of our floating community had greatly increased, and pleasant conversation, music, chess-playing and such like, made the time pass quickly.

There were on board forty first-class passengers, amongst whom some had a strongly marked individuality, with boldly-outlined, well-coloured lives. Fresh, hopeful young Englishmen, going out to ostrich

farming and diamond-digging; old colonists bringing
home their daughters from English boarding-schools;
a clergyman returning from a holiday won after sixteen years' labour on St. Helena; a young German
girl from the Hartz Mountains on her way to a mission
station in Riversdale; and two American ladies, also
destined for mission work in Natal.

In conversation with these ladies and a life-long
colonist, the son of a South African missionary, I
learned much of the mission work among the Kaffirs;
of its progress, and alas! of its many discouragements
and difficulties. It is hard to know how Christ can
best be shown to these dark brothers of ours. So
much care is needed in preparing the uncultured soil
for the good seed, that the poor missionary must have
fear and trembling in dealing with the truth. Careful
on the one hand to avoid compromise with its perfect
purity, and on the other, fearful of driving back any
wavering disciple by making it too hard for him to
receive.

Our colonial friend, a man of culture and common
sense, seemed to think that many of our most zealous
missionaries fail from a want of true knowledge of the
mind and nature of the savage. They would have the
slaves of ignorance and superstition accept eagerly the
freedom offered in the Gospel, forgetting that the soul
must first be strained to the note before it can vibrate
to the sound of the glad tidings. It must be a work of
time. To forge a single link in the golden chain that
is being formed to draw the poor heathen to the feet

of his Maker, is surely a work large enough for life—a
work complete enough for death.

I was so much interested in hearing the experience
of one who had seen so much of heathen life and
Christian teaching, and in being told of the hopes and
plans of those who had left home and friends to make
this life better and this teaching more perfect, that I
was sadly rebellious when, after a truce of three days,
my old foe again forced me to beat a retreat to my
cabin. There I had generally nothing more lively to
listen to than the irritating *thud* of rope-quoits and
shovel-board.

But sometimes I was better amused. Stray threads
of conversation, grave and gay, floated through my
little window at intervals as the passengers paced the
deck, and in the lovely moonlight evenings, a young
couple, thinking no doubt that they had got away into
a nice quiet corner all by themselves, used to coo their
tender speeches unreservedly, close to my cabin door.
It was mean to listen, I grant, but ennui was threatening me, and during these enervating days, when a damp
monsoon blew off the coast of Africa, I was so nearly
reduced to a pulp, that I had hardly energy left to
make my presence known.

Then, for the first time, I thought kindly of our
east winds, which I believe have done much to make
England a great nation. How I should now have enjoyed
a biting blast from Scotland! Wearied of inaction, I
often caught myself counting off the days and hours
like a home-sick schoolgirl; and yet there was no

delay to chafe the most impatient voyager. On and on, the great ship rushed through the waters ; the bells struck the passing hours, and every noon the answer to the anxious question, "What's the run?" told of nearly 300 miles further on our way. At last on the 1st of July, at 4 A.M., the screw suddenly stopped, and I knew and rejoiced that we were in the Bay of James Town.

At the first peep of dawn I hurried on deck and saw, so close to the ship as to make me start, dark sterile rocks rising almost perpendicularly from the sea, and partly enclosing the bright blue bay in which we were anchored. At the bottom of a strange cleft in these fierce, fortress-looking crags, a quiet little town nestled close to the sea, filling up the lap of a valley scarce 200 yards wide.

Here was the landing-stage, and just beyond, a row of dark Peepul trees fringed the shore, shading and cooling the cluster of low, white houses that we were so blithe to see. Besides these, little or no vegetation appeared. The great towering rocks were cold and bare. A long ladder of 600 steps sprang from the town up the steep western side, called Ladder Hill, and at the top I thought I could descry some forts and the grim mouths of cannon.

St. Helena can hardly be mentioned, much less looked upon, without memories of Napoleon crowding upon us, and I wondered, as I suppose everybody does on seeing the island, how the first sight of these grim prison walls had affected the man who seemed to find the world too small for him ?

But this was no time to speculate on questions of by-gone history. The present was urgent, and suddenly remembering the terrible chaos in my cabin, and boxes still unpacked, I gave up dreaming and set to work. How very easy it is to pull things out of a box, and how difficult to get them into it again, especially in a space 7 feet long by 3 broad!

The Governor's Secretary was already on board, having kindly come off thus early to advise David about landing and stowing away his numerous cases. At Lord Lindsay's kind instigation, Lord Carnarvon had previously sent despatches, requesting assistance for him in this and other matters ; and for the timely help thus given we were most grateful.

That lovely Sunday morning the bay was smooth and bright as a mirror, without a trace of the dreaded rollers, so that we and all our delicate impedimenta came safely and easily to shore. The Governor's pony carriage was on the wharf, and while David counted and sorted out his goods, I was driven up a short incline to the Castle, where I waited for him comfortably in the large cool rooms.

Here I occupied myself in watching for the return of four gentlemen, our fellow-passengers, who had set out soon after daybreak for Napoleon's tomb, in the interior of the island. They were still missing, although the *Balmoral Castle* was getting up steam; and the Captain, kind as he was, would be off the moment the ship was ready to sail. That moment arrived; and just as the ship's bows were disappearing round the rocky headland, a single figure rushed frantically upon the pier, and next minute a white handkerchief was floating from an oar. Would the good ship see this flag of distress and put back? Yes. With the help of a telescope, I was watching events from an upper window in the Castle, and rejoiced when this " derelict " was saved.

But there were still three unlucky ones to be sorry for. And, as first the bows, gradually the long white hull, and finally the Union Jack on the stern disappeared, I felt that their case was hopeless; and so it proved. By-and-by other hurrying figures were seen to pull off in pursuit, but only to return disappointed, doomed to five weeks' captivity on St. Helena. We sympathized deeply with them, especially after we had seen the " Imperial Arms," where they must take up their abode, and noted the dull, dead-looking street which forms James Town. A street of rickety, blistered houses and of dusty ant-eaten shops, with untidy and untempting goods therein displayed, and closed in, almost to suffocation, by rocks on either side.

The ground rises quickly from the shore, and as we drove slowly along, a curious mixture of faces crowded at the open doors and windows. All shades were there, from the woolly, jet-black Hottentot to the fair-complexioned English sailor, leaning against the door-post of the " Royal Banner," with " H.M.S. Cygnet "

on the ribbon round his cap. After a short stiff pull, we
left this motley crew behind, and a shady lane soon led
us into a lovely garden where palm and pomegranate
trees shaded the rich luxuriance of sweet-smelling
roses and scarlet geraniums. A low-roofed verandahed cottage formed the centre of this little Eden,
and here we found a comfortable home during our
stay.

No sooner was our baggage safely landed, than
David began to make inquiries about the Observatory
on Ladder Hill and the best mode of access to it. It
was in this Observatory that Johnson, about fifty
years ago, made his catalogue of southern stars; and as
its longitude had been determined with considerable
accuracy by that astronomer, my husband was anxious
to connect it by means of his chronometers with
Ascension, and thus strengthen the results for the
longitude, which he might get at Ascension, from
observations of the moon. For this purpose it was
necessary that the chronometers should be conveyed
to the Observatory, that local time should be determined there morning and evening by observations
with the reflecting circle, and that careful comparison of the chronometers should be made every
day.

Ladder Hill rises about 600 feet above sea level
and forms the western side of the valley in which
James Town lies. There are two ways of getting
to the top—the one by means of the ladder I have
already mentioned, the other by a very zig-zag carriage-

road, which winds along the side of the hill. It was by the latter way that the instruments were conveyed by some gunners, through the kindness of Captain Oliver, R.A.; and David's next business was to hire a strong little Cape horse to carry him up the same road every morning and evening on his visits to the Observatory.

I say Observatory—alas! it is so no longer. Fallen from its high estate, it is now the artillery mess-room, and in the recesses formed for the shutters of the openings through which Johnson's transit used to peep, they stow wineglasses and decanters, and under the dome they play billiards! It may appear un-grateful to speak so of a change which was produc-tive of so much kindness and hospitality to us; I do not grudge the hospitable St. Helena Mess their mess-room, but I do regret that so fine a site for an Observatory is vacant.

Another kind friend and sympathizer in his work, my husband found in the Governor, Mr. Janisch. An enthusiastic amateur in astronomy, a descendant of the great astronomer Encke, born on the island and spending his whole life there, he had never before met an astronomer, and the welcome he gave was warm and cordial.

Indeed Mr. Janisch had so much to urge in favour of St. Helena as an observing station for Mars, that he had almost tempted us to remain; and when we saw the clear cloudless sky, night after night begemmed with stars, there did creep into our minds a doubt of meteorological statistics, and a fear lest in going further we might fare worse.   But the risk must not be run.   Had all succeeded here, well and good, and probably my husband would have been congratulated on his change of plan, but had bad weather come and failure resulted, there would always have remained the reflection—" Why didn't I go to Ascension ? "

CHAPTER III.

ST. HELENA.

BEFORE entering upon the work that we had left
England to do, it was a kind chance that took us for a
holiday to St. Helena. Tired of anxious preparation
and the constant thinking of one thought, it was
no real loss of time to turn aside for a rest by
the way, and gather fresh strength from fresh scenes
and from the most delightful air it is possible to
imagine.

During our week's stay we were able to make three
excursions on horseback from James Town, and these
little peeps into the country showed me so much that
was strange and new, that I find it difficult to dis-
entangle one impression from another. So quickly
did they follow in succession, that the one partly
effaced the other before time had allowed it to harden

on the mind, and the picture memory has to show is somewhat blotted and confused.

The sun had been shining some hours on the hill tops, and was just beginning to creep down the dark rocks into the valley when we started for our first excursion.   Captain Oliver had kindly offered to be our guide to Diana's Peak, the hill *par excellence* of St. Helena, and towards it we now wended our way southwards and upwards.   A narrow bridle-path, curling itself among aloes and wild geraniums, soon brought us to " The Briars," where Napoleon spent the first month of his imprisonment.

It is a pleasantly situated house, within view of the sea at one point, but on all other sides so shut in by steep rocks that I wondered how we were to proceed. The path, however, gradually opened up as we wound round the hills, and new beauties burst upon us at every turning.   Now a sudden bird's-eye view of the little town, looking like a cluster of card houses far below ; now a ribbon of clear water breaking into feathery spray as it falls over the steep cliff; here and there a group of goats or shaggy calves giving life to the picture ; and on every bit of level ground some pretty white villa with its trim garden relieving the wildness of the scene.   Straight-limbed aloes were shaking their feathery blossoms thirty feet overhead, and homelier flowers were creeping humbly round their stems. The snow-white blossoms of the beautiful moon-plant (*Brugmansia suarcolens*) scented the air. The Hottentot fig (*Mesembryanthemum edule*) trailed its delicate starry flowers among the dark leaves of the yam. The knarled cabbage-tree (*Aster gummiferus*), the castor-oil tree, Port Jackson willows and graceful acacias mingled their shades of green with the deep red blossoms of the wild fuchsia and the pale yellow of the rock-rose (*Hibiscus arenatus*).

Winding along the upward path, we came to a gap in the ridge of hill through which burst upon us a glorious view of the southern coast. For the first time in my life I saw the naked grandeur of volcanic action, undimmed by time, unsoftened by vegetation. Wild grotesque masses of rock, assuming every shape and aspect, were piled up, one above the other, with fantastic irregularity, and tinted like the opal by the noonday sun. Beyond lay the sea, of a deep blue where it bordered the rocks, but gradually becoming softened in colour by the shadow of a great white cloud far away in the horizon. The scene was altogether so unlike anything I had ever conceived to be among the beauties of the world, that I could have imagined myself somewhere in space and looking down on one of those gorgeous, fairy clouds that we sometimes see floating in the summer sky after a thunderstorm—masses of colour, gloriously bright with the brightness of the setting sun, then soft and tender as he bids them farewell, and finally dark and sad-hued, mourning the death of the great painter.

Such is Sandy Bay on the south coast of St. Helena, or rather, such is it as I am able to describe

it. A rift in the ridge had shown us this glorious
picture ; presently the rock towered above us again, and
we saw it no more. Indeed I saw little landscape of
any kind about this point, for the path had got so *ugly*
that I shut my eyes and ignominiously grasped the
pommel of my saddle. To make matters worse, I had
been told that the pony I rode was a "buck-jumper,"
and I could not help wondering what would be the consequence should he take it into his head to exhibit any
of his feats just at this moment. But this bit of
nervous riding was short, and fortunately my courage
was able to hold out until we came upon a more
level road. About a hundred yards below the Peak
we tied our ponies to one of the numerous gates that
are placed along the pathway to prevent the straying of
cattle, and climbed on foot to the top.

Here we could command at a glance the entire
length and breadth of the little island, as well as the
unbroken circle of sea surrounding it. To the north
lay James Town, hidden in its narrow valley ; to the
south, Sandy Bay with its chaos of wonderful rocks
throwing out their grand outlines against sea and sky.
One of these detached rocks is no less than 1,444
feet in height. It is an upright precipitous mass
of greenish grey phonolite, known by the name of
"Lot," and "Lot's Wife" stands near him, a fit mate
in size and beauty. To the eastward we looked down
on a gentler bit of landscape. Here lies the only
large tract of cultivated land on St. Helena ; and the
square green fields, farmhouses, and little church

perched on a wooded knoll in the background, took
us in fancy back to England.

It was towards this side of the island that we
began our descent from the Peak, through fuchsias,
blackberries and ferns of all kinds—from the gigantic
tree-fern to the tiny *Acrostichum bifurcatum*, creeping
shyly into nooks, as if it were ashamed of the big
name which botanists have given to it. In one or two
places we noticed the curious grass-like *Polypodium
marginellum* growing parasitically on the tree-fern, and
here too, choked, alas! and laid waste, are still to be
seen some plants of *cinchona*, which our Government
began to cultivate on St. Helena by the advice of Sir
Joseph Hooker. The plants flourished well while
care was given to them; but this is no longer done—
a neglect much to be deplored.

The downward road was more level and less fatiguing than the path we had followed in coming up.
Nevertheless we were glad to dismount for rest and
refreshment at the " Rose and Crown," the only house
of entertainment on the island, except, of course, the
hotels of James Town; and, together with its landlord, it
was by no means the least curious thing we had met with
in our day's ride. Tom Timm, his dusky face aglow
with heat, and the extraordinary excitement of three
guests all in one day, rushed out, napkin on arm, with
the welcome greeting that luncheon was ready. A
long, uncarpeted, unceiled room was the salle-à-
manger, with bunches of stags'-moss adorning the bare
rafters, and on the walls were many works of art,

dark and mysterious-looking enough to be "Old Masters."

But Tom himself was bright and by no means mysterious. He most good-naturedly entertained me with his stock of local gossip, while Captain Oliver and David strolled along to "Halley's Mount" to search for the site of the Observatory where Halley, in 1677, made his catalogue of Southern stars and observed the Transit of Mercury. We did not know whether any record of this work remained in stone and lime, and it was a pleasant surprise to find, on the spot that an astronomer's eye at once picked out as the most favourable, a bit of low wall, duly oriented, and overrun with wild pepper (*Cluytia pulchella*). This had been the Observatory, without doubt; and near to it is a quarry from which the stones for its erection had evidently been taken. So charmed was my husband with this interesting record of the work of 200 years ago, that his investigations and surmises regarding it left us short time to linger in the little hollow lying near the foot of Halley's Mount.

Napoleon's tomb is here. It is a lovely spot that the great General chose for his last resting-place, close by the clear spring that used so often to refresh him after his walk from Longwood, over a mile distant. We found the place under charge of a French sergeant, and almost over-trim in its exquisite neatness. A plain iron railing encloses a plot of mossy grass, shaded by cypress, willow and other sombre trees, and an inner rail, round which climb bright geraniums, pro-

D

tects the tomb itself. An ancient-looking, leafless willow hangs over it, but this is not the original willow as I had fondly hoped. *That* has been ruthlessly hacked to pieces long since by relic-hunters, and this lineal descendant, though better protected, already looks tattered and forlorn, and will, no doubt, soon die the death of its predecessor. With a view to this fate indeed, a younger willow has been planted close by to take the place of honour when the present tree falls.

Relic-hunters are the Goths of the age, and something of their savage nature must be in me, for there was no resisting the impulse to gather a few leaves from the geraniums which wreath the empty tomb. St. Helena is so rich in associations of Napoleon the Great, that one breathes them in with the air, and infected for the time with the insanity of hero-worship, we can hardly escape the relic mania.

Another day we visited Longwood Old House, where the term of Napoleon's imprisonment was spent, and where he died. The house stands in the interior of the island, on a somewhat bleak and treeless plateau, 2000 feet above the sea. To reach it, we followed the carriage-road which winds up the eastern slope of James Town Valley. Now north, now south, this corkscrew road led us, now facing the sea, then turning to the land—so that I lost all idea of direction. But I had confidence in our ten-year-old guide, who kept pace with us by twisting his hand into my pony's tail and so pulling himself along,—a universal

practice, which says much for the ingenuity of the
" Yam Stalks " (St. Helena natives), and for the good
temper of their horses.

Arrived at Longwood, we left the ponies under
charge of our guide, and opening a little wicket, we
walked through a short garden-path to the door of
the low rambling house, where a sad-faced woman
received us politely and conducted us over the different rooms, telling us what had been the use of
each. The English and French flags are crossed
over the fireplace in the entrance hall or "salon à
fumer," and the room contains nothing besides.
Immediately beyond is the room where Napoleon
died, its only ornament being a laurel-crowned marble
bust, standing on the spot where he breathed his
last. All the rooms are in good repair, but unfur-
nished, and smelling of disuse.

The woman in charge told us that formerly all
French sailors visiting St. Helena used to be marched
up to Longwood House, but the place so excited their
quick imaginations that they became quite wild, and
they have now been prohibited from visiting it. They
exhibited their enthusiasm chiefly by chipping pieces
from the door-posts and stripping the paper from the
walls. Nor have English travellers been guiltless of
aiding in this work of destruction.

Mellis says, " A remarkable instance occurred of this
bad habit of relic-stealing being turned to good account.
It was wished that the rooms might be made to
look as much as possible like what they had been

when occupied by Napoleon; but a great difficulty
arose about the wall-papers. Not a scrap nor a vestige of them remained, and no clue could be obtained
as to their design or colour. This difficulty reached
the ears of an English officer who had visited Longwood thirty years before and carried off a scrap of
paper from each room. These specimens, which had
been carefully preserved, he at once placed at the disposal of the French engineer in charge of the work,
who sent them to Paris, where new papers exactly
resembling the originals were manufactured and sent
out to St. Helena."

Close by Longwood Old House stands Longwood
New House, built for Napoleon, but never occupied
by him. It was not completed until shortly before his
death, and he refused to move into it, notwithstanding
its superior accommodation. In the same cluster of
buildings there is also the cottage which was occupied
by Marshal Bertrand during his attendance on Napoleon. It now serves as the dwelling-house of
Longwood Farm, which we had already admired
from Diana's Peak. Having previously received a
kind intimation that the farmer would gladly show
whatever might be of interest to us, we now took
advantage of this proffer to ride across the fields with
him, and see the different agricultural operations that
were going on. From this and the neighbouring farm,
" Teutonic Hall," come the chief supplies of James
Town. This is due to the energy and skill of two
English farmers, who, with their families, have turned

to good account the rich soil of decomposed lava,
which is ready and willing to yield food for man and
beast.

But farming in St. Helena, as well as farming at
home, has many drawbacks to contend against. The
last crop of potatoes had been entirely lost through
want of rain, not enough having been saved for seed,
which had to be brought from England at great cost.
Then, worst of all, is the want of a market for cattle.
Since Christmas, fat cattle had been ready at Longwood Farm for shipment, and no ship had come to take
them. All this must be taken into account as well as
the very high price of labour; 2s. 6d. a day being the
usual hire of a farm labourer, and the "Yam Stalks"
do not work with the energy of Englishmen. But
they are obedient, and, once set a going, go steadily on
like machines. In Scotland we should characterize
them as " eident "—untranslateable into English; but
" slowly and surely industrious " gives some idea of the
meaning.

Here for the first time we saw the light-coloured
island partridges flying over the garden-like fields,
which are separated from each other by hedges of
cacti and scarlet geraniums. How gay it was! The
bright sunshine, the bright flowers and fields, the
golden-winged canaries flitting hither and thither,
darting in and out of the hedgerows, their sweet notes
almost drowned in the husky whirr of the grasshoppers.

Our third excursion was made, as the first had been,

under guidance of Captain Oliver, and with the pleasant addition of another artillery officer and his wife. This time our guide led us into the western and most beautiful part of the island. Another cork-screw road drew us slowly to the top of Ladder Hill, and then we cantered pleasantly along by Friar's Valley—so called from a curious rock of dark basalt here, which is supposed to resemble a cloaked and hooded friar, who suffered as a renegade on the spot where it stands.

The legend tells us that : " The place where the Friar Rock now stands was once the site of a church, adjoining which was the residence of the officiating priest, who was looked upon as a model of Christian piety, passing his life in acts of charity and benevolence. Blessing and blessed, this man of God pursued his way, until he allowed himself to be enthralled by the wonderful beauty of a mountain girl who dwelt near his home. It was in one of his rambles on some charitable mission, that the ill-fated friar first saw this lovely shepherdess tending her father's goats on the adjacent hill, now called ' Goat Pound Ridge.' They had strayed so far that she had vainly tried to collect them and was returning home, tired and sad, when she met the monk, to whom she told her tale and begged his assistance. It was given, and the scattered flocks soon collected, but more evil than good was done. It would have been well for the good friar if this meeting had been the last, but fate ordained it otherwise.

"Again and again he sought the mountain hut with a tale of love, and finally besought the maiden to be his bride. She promised, but on one condition—he must renounce his creed and become of her faith. The struggle was a strong and fearful one in the heart of the monk, but—

'Love must still be lord of all.'

"He forsook the faith of his fathers, broke his vows and became a renegade. In the course of time the wedding-day arrived; the bride, accompanied by her attendant maidens, had approached the altar, the ceremony was proceeding, and just as the bridegroom was clasping the hand of his beloved, a fearful crash resounded, the rock was rent asunder, and every vestige of the chapel and of those within it disappeared for ever, leaving in its place the gaunt figure of the grim friar. A warning, says the moral, to those who suffer passion to stifle conscience."

Such is the story of the unhappy monk—I wonder what geologists think of it!

The surface of this part of the island reminded me somewhat of a honey-comb, into the cells of which we now and again descended, finding always at the bottom some pretty villa, nestling among acacias, or a white farmhouse standing in fields black with rich mould washed from the encircling hills. Sometimes our road left the cells below and wound along their turf-covered ridges, thus allowing us to obtain a fairly good idea of the general topography of the country.

I have the vaguest notion of how many miles we might have ridden along this zig-zag, up-hill-and-down-dale road. I only know that after three or four hours of it, I did not object to halt at West Lodge for our pic-nic luncheon, which Captain Oliver, with kind forethought had despatched, donkey-borne, early in the morning.

A gloomy, half-ruined and haunted house is West Lodge, but all around it is bright, and smiling vistas of wooded knolls and flower-clad dales stretch far away among the hills. Beautiful ferns embellish every nook of the half-wild garden, and here and there along the paths are stationed great camellia trees with a stately burden of crimson and white flowers.

But this was only one of the many pretty country residences which we observed tenantless and in a state of ruin. Naturally these signs of a decreasing population made us look about for an explanation, and several reasons presented themselves.

Formerly almost all vessels coming from the East called at St. Helena for fresh provisions, &c., and it might be reckoned that a thousand ships a year, in former times, cast anchor in James Bay. But now they make swifter passages, and can easily accomplish a voyage from the East to Europe without an intermediate stoppage. This, with the opening of the Suez Canal, has reduced the number of ships calling at St. Helena by nearly one-half.

Then the garrison is greatly reduced, and many of the civil offices have been abolished, the line of policy pursued by the Home Government towards St. Helena being characterized by a somewhat ruthless economy. Plantation House is the official residence of the Governor, but he finds it more convenient to occupy his private house in James Town, and owing to reduced salaries on all sides, even Plantation House has not quite escaped the infection of general decay. But nothing can rob it of its beautiful surroundings. A square compact mass of building, of no architectural pretensions, it stands facing a beautiful park, dotted with groups of trees of innumerable shapes and shades of colour. Below the park are the famous gardens, containing fruit trees and tropical and sub-tropical plants in such wonderful variety, that all our time in St. Helena, and more, would have been needed to examine them thoroughly.

On a rising ground behind Plantation House, the little cathedral of the island peeps from amid a grove of magnificent cypress trees, which dwarf its tiny spire, and, with their sombre masses dark against the pale blue sky, form a perfect background to the view as we saw it, riding home from West Lodge in the twilight.

But perhaps the most beautiful of all St. Helena's beautiful homes is Oakbank, the residence of the Bishop. I cannot recall a more lovely spot. Nature seems to have denied nothing to this pet child of hers, and Art has helped her gracefully, controlling without thwarting her. When we were there, the oak trees, from which it takes its name, were leafless, and their naked arms, interlacing with the bright green boughs of neighbour trees which acknowledge no winter, dashed the forest picture with great streaks of grey; and the rustle of withered leaves under our horses' hoofs was a homely, autumnly sound.

Oakbank is the queen of pocket landscapes, but in every gully here, little gems lie hid that would delight a painter's eye, and the variety of scenery within so small a compass is indeed wonderful. Grand rugged rocks, gentle, grassy slopes, tilled fields and hedgerows, gardens of palms and pomegranates, beds of violets and mignonette, clumps of pine trees, waysides of gorse, and everywhere the sea. All this St. Helena showed us in a week. No wonder then that we found it a happy one, and that we brought away with us bright memories to think and talk over among the barren rocks at Ascension.

On Monday the 9th, the Cape steamer was due, and, learning caution from the fate of our laggard fellow-passengers on board the *Balmoral Castle*, we held ourselves in readiness from daybreak. These unfortunate gentlemen we had met with several times during our ramblings, and they really seemed to bear their misfortunes bravely, making good use of their unexpected time on St. Helena.

We subsequently heard that the one whose business at the Cape was the most pressing had been taken on board a troop-ship, that called shortly after we left. The Captain would not be induced to take the others, owing to the already crowded state of the ship; but one of them, careless of all consequences, surreptitiously stowed himself on board. The third, from feelings of self-respect, decided not to have recourse to this plan, and he probably fared all the better for his decision. Just as the troop-ship was under way, the mail-steamer from the Cape, bringing the various effects of these unfortunates, entered the harbour, and thus missed two of them by a few minutes. The one who had remained behind, no doubt felt his virtue rewarded, and so charmed was he with St. Helena, now that his purse and wardrobe had been restored to him, that he resolved to enjoy himself there for a few weeks longer.

It was the morning of the 10th before the call came for us. At 7 A.M. the *Edinburgh Castle* was signalled, and some hours later we went on board, accompanied by a large party of the kind friends who had given us such warm welcome to St. Helena, and whose hospitality had added so much to the pleasure of our visit. We were loth to say good-bye. From the Governor we parted with great regret, and we shall always retain the strongest feeling of gratitude for the sympathy and assistance he gave us in our work. Certainly while Mr. Janisch is Governor of St. Helena, any astronomer visiting the island will find a zealous supporter and a kind friend.

With so many St. Helena friends on board we did not feel as if we had quite said "good-bye," till a noisy, impatient bell rang for the third time. Then last handshakings were given, hats and handkerchiefs waved, and as little boats pushed back to the wharf, we steamed into wider waters, gradually losing sight of those "grey beetling crags" which hide so much softness and beauty. No thunderbolts nor lightning shafts, no burning drought nor deadly disease, no savage brute nor noxious reptile, not even *a lawyer;* surely this St. Helena, now melting away in the distance, must be the "The Island of the Blessed" so fondly believed in and so earnestly sought for by the ancient mariners.

CHAPTER IV.

WHAT ASCENSION LOOKED LIKE.

Clarence Bay.—The "Abomination of Desolation."—Is that Ascension?
—Rollers.— Can we land?—An uninviting dingey.—A slippery
footing on Ascension.—A kindly welcome.—"The Thing."—The
beauty of perfect ugliness.—The Captain's Cottage.—Between
hammer and anvil.—Commodore's Cottage.—A bare larder.—
Shopping in Ascension.—Threatened starvation.—A novel Croquet
lawn.—Site for the Observatory.—Glorious skies.

THREE days brought us within sight of Ascension.
What a sight it was! The sun had been up some
hours when we anchored in Clarence Bay on the 13th
of July, and the "Abomination of Desolation"
seemed to be before our eyes as we looked eagerly at
the land.

A few scattered buildings lay among reddish-brown
cinders near the shore—a sugar-loaf hill of the same
colour rose up behind and bounded the view. We
looked about in a sort of hopeless way for "Green
Mountain," but it was nowhere to be seen, and we set
it down as a fable—a mere myth. "Nothing green,"
we said, "exists, or could exist here." Stones, stones,
everywhere stones, that have been tried in the fire and
are now heaped about in dire confusion, or beaten
into dust which we see dancing in pillars before the

wind. Dust, sunshine, and cinders, and low yellow houses frizzling in it all !

Is *that* Ascension ?

Well, not quite ; its coast presented a livelier scene, though one that we would gladly have dispensed with. A black perpendicular wall of rock jutted out into the bay, and on either side of it a stretch of white glistening sand swept to north and south. It is on this rock that the " Tartar Stairs " are cut, and here we must land. But how ? For this morning beautiful waves are dashing and crashing and splashing against the landing-place, or rushing past it in sportive fury to break into feathery foam on the pretty beach, which looks like a dainty white ribbon trampled under foot of these mad sea-monsters.

" The rollers are in ! " " What lovely waves ! " " What a hideous place ! " were the ejaculatory remarks we heard drop from the ladies leaning over the ship's side. My heart grew heavy. But seeing H.M.SS. *Cygnet* and *Industry* in the harbour, I took courage, knowing that we should at least find refuge on board one of these vessels, and that we should not have to be carried on to Madeira,—a misfortune which has more than once happened to passengers roller-stayed at Ascension.

There were besides several little heaving boats in the bay, and one could not but wonder at their audacity in playing so unconcernedly with the mighty giants that tossed them about, each in turn, as one after one rushed headlong to the shore. While watch-

ing this scene, we saw a gig put off from the *Cygnet*,
and pull towards us. " An offer of hospitality," we
thought, as we recognised the blue-jacketed oarsmen
and their commander, whose acquaintance we had
made at St. Helena.

" Can we land ? " was our greeting to Capt. Hammick, as he came on board. " Well, the flags denoting
' Double-rollers and Dangerous ' are up on the pier-
head, but the sea is going down, and I have permission for you to try it, if you don't mind wet feet."
We didn't ; so it was decided that I and the heavy
baggage should be sent on shore at once, while the
chronometers and more precious goods ! should wait for
quieter times on board the *Industry*, where the Captain,
in the kindest manner, had prepared his cabin for us
in anticipation of our not being able to land.

I don't know how the heavy baggage liked it, but I
certainly wished myself a chronometer more than once,
when I saw, rising up behind us, a long wall of threatening water, and before us, the steep, dark rock, wet
with spray. This feeling increased when we were
within a few yards of the shore, and I found that we
must get out of the strong trustworthy-looking gig,
manned by its stout crew of English sailors, and trust
ourselves to a little rickety cockle-shell, which was at
that moment being baled out by two ebony-coloured
boatmen. I thought, just then, they looked fiendish,
and that I could see the baleful eye of a shark, certain
of his prey, gleaming triumphantly through the green
waves. But since then I have come to the conclusion

that our boatmen were very benign, gentle-faced Africans, and my shark—a jelly-fish!

"You may trust yourself with every confidence to these men," Capt. Hammick said to me; "they understand the rollers better than anybody else; they will not take you into danger, only you must be careful not to attempt landing until they give you the word."

For some minutes we kept dodging about, and once or twice were close under the steps; but we got no sign to stir, and were again and again driven back.

At last, there came suddenly a perfectly calm moment, immediately after an unusually heavy roller had tossed our little boat over its head, and we were again sculled under the rock in the twinkling of an eye. A rope was let down from above; David at once laid hold of it, and at the word "Now!" he jumped from the boat. I instantly followed his example, and thus gained a slippery footing on Ascension, with a somewhat palpitating heart and eyes smarting with salt spray.

Among a little group of officers and men on the wharf, we found Capt. Phillimore (the naval officer in command) waiting to welcome us. He very kindly offered us the hospitality of his house until our own cottage should be made comfortable; so, while David braved the rollers a second time to make sure that all his goods were off the steamer, I gladly accompanied the Captain in a curious two-wheeled vehicle—which my conscience would not allow me to call a carriage, and which I was afraid to call a cart, lest by so doing

I might commit some breach of etiquette. Not knowing the manners and customs of the natives, I felt the
safer plan might be to call our means of locomotion
" the thing," in faithful imitation of Miss O'Dowd's
coachman of comic memory.

It was now nearly noon, and the dazzling sun shone
with a pitiless glare on everything. I looked about
me for some beauty to remark upon. But no! We
passed great open sheds, piled roof-high with coals,
square unsightly store-houses of various kinds, a creak-
ing windmill painted red like a guillotine, and all
thickly coated with a fine yellowish dust, into which
our poor horse was sinking, hoof-deep, at every step as
he pulled us up the gently rising ground leading from
the wharf.

Having surmounted this we came upon a dreary flat,
and still dust and ashes everywhere. Here we found
facing us a neat little church; to the right the hospitals
and marine barracks, with their two stories, interrupted a row of low-roofed, verandahed cottages, one of
which I gladly learned was to be our home. Beyond,
were a few scattered, undecided-looking houses, with
no character to speak of. We drove through these
before beginning the ascent of Cross Hill, now rising
straight before us, with the Captain's Cottage about
half-way up the steep slope. Again dust, ashes,
cinders; and paradoxical as it may seem, this was a
hill without beauty, except perhaps that beauty
which characterizes the fashionable lady's pug-dog—
the beauty of perfect ugliness.

E

Without a tinge of green, without a single rough
shoulder to catch the sun and throw a shadow, this
was no *real* hill, but simply 800 feet of smooth cone-
shaped cinder-heap, surrounded by smaller cinder-
heaps at irregular distances; and to render the scene
still more ghastly, some white grave-stones peered
through the cinders by the side of the road.

Very glad indeed was I to leave this miserable prospect, and to get into the pleasant shade of Captain
Phillimore's verandah. Here I was met with that
warmth of welcome which is without doubt the most
effectual rest after the fatigue of a journey; and this,
followed by the reviving influences of a cup of tea,
fortified me for the depressing information which I now
received as to the limited resources of the island.

" Scarcity of food and servants!" Well! that was
not very encouraging; but fortunately, just at present
there happened to be on the island two convalescents
fit for work,—one white, the other black,—both
invalided from service on the Gold Coast. We seemed
to be in luck; but alas! when these "gentlemen
helps" came to be interviewed, it appeared that each
wished to be master—one had been a ward-room cook,
the other a steward—neither would scrub floors nor
run messages, so, fearing lest I should be between
hammer and anvil with so much talent in the house, I
resolved to be my own steward, and engaged only the
white cook. He promised to wait at table; and
Captain Phillimore kindly undertook to let me have a
Krooman to do the "low caste" work.

My ménage thus far arranged, I felt "settling down;" and when David joined us, he brought the further good news that the rollers were "settling down" too, and that probably all our goods would be on shore before sunset. This being the case, we thought it better to lose no time in taking possession, and after an hour or two's chat we returned to George Town, as maps and geography-books call the little settlement. In Ascension itself it is called "Garrison," and we soon knew it by no other name.

I did not find that things improved on closer inspection; for now, in walking down the hill, the hot cinders burned through my thin boots, and I looked eagerly about for the neat square gardens and paved streets seen by Sir Wyville Thomson. But I could make out only a few tortuous paths of concrete leading to the chief buildings, and along the back of a single line of gardenless cottages, one of which had "Commodore" painted in white letters on the verandah gate. It was of this one that we now took possession. On either side of the door were placed the divided halves of a cask, painted green, and containing what *ought* to have been a green shrub. These floral ornaments were the nearest approach to gardens that I ever saw in "Garrison," and they could never, by any stretch of the imagination, be called "square," though, as an Ascension lady very wittily suggested, "tubular gardens would not have been amiss."

Of course, I explored every nook of my little home,

E 2

before attempting to fit myself into it : and though I was
delighted with its dimensions, I cannot say so much
for its contents.   Entering by a glass-door on the
north-west, I found on the right, a dining-room and
drawing-room, both of good shape and size, and opening into each other.   To the left was a tolerably large
bed-room, with dressing and bath-rooms beyond.   The
little kitchen was built behind the dining-room—the
only case in which I saw this arrangement in Ascension—the kitchens or "galleys" being as a rule
separated from the rest of the house to avoid the heat
of the cooking-fire.

Commodore's Cottage stood, like its fellows, facing
the sea and the north-west, with green jalousie doors
to the back and front of each room, so as to give free
course to the refreshing trade wind from the south-
east.   This was all very well if the wind had not been
troublesome as well as refreshing.   But how cross
it made one, after a severe fit of tidiness, to find
newspapers, pamphlets, writing materials, and such
like, strewn about the room in wild disorder!   Yet it
tried the temper more to sit stifled with the glass shut
on the " win'ard " doors ; so I preferred the jalousies ;
and notwithstanding that my watch, card-case, and
everything available were utilized as weights, it cost
me many a chase after stray papers, indoors and out in
the verandah which surrounded our Garrison home on
all sides.

But to return to the interior, as I found it on the
night of our arrival.   I have said that I was not

delighted with the contents, but in one sense I was. I have always thought how very tame and uninteresting it must be to take possession of a house where the upholsterer has done everything; where every detail is perfect, and every nook filled, even the book-shelves in the library. I should feel tempted to turn back at the threshold, fearing to disturb, even by a restless thought, this "faultily faultless" establishment. So it was, that I gloried here over bare walls, bare floors and bare tables; till I was disturbed in the delightful occupation of mentally putting my house in order by Hill, our brisk cook, who came to remind me that the larder was bare too.

Then I gave up my castle-building, and accompanied Hill to buy and lay in provisions. But this was by no means so simple a process as I had expected. No butcher! no dairy! no greengrocer! no fishmonger! only this wretched canteen, more full of flies than of anything else. I got quite tired and hot with the frequent, "No, madam, we don't keep it," or "Very sorry, but we are just sold out." My demands were modest, but they had to become yet more humble before accommodating themselves to the limited resources of the "Royal Naval Canteen." Finally, however, I succeeded in getting some sustenance for the body.

I then turned to the open door with "Island Bakery" written over it—where a pallid baker stood at the threshold wiping the perspiration from his forehead. Evidently *he* made his bread by the sweat of his brow!

"Can I have some bread?" I asked boldly, thinking there could be no difficulty here. "All served out for the night, ma'am." "Oh dear! and when do you bake more?" "The day after to-morrow!" and my heart was sinking, when the good-natured baker added "But I can make you a loaf now, if you like." Then I revived.

Now about milk—which David and I were wont to consider a necessary of life. I was told, "a mule brings that down every morning from Green Mountain, *when there is any.* A bell rings at 7 o'clock and everybody runs for a *gill*, except when there are many sick in hospital, then *they* get it all." This was lively! "And vegetables?" "There are only sweet potatoes to be had, and none will be 'served out' until next Friday."

Then came the most important question of all. "Where shall I find the butcher?" "Oh!" said Hill, with a grin, "there ain't any butcher. One of the marines kills sheep twice a week, and on Saturdays a bullock, which is 'rationed' out, so much to each man, and our rations are very small just now, for the sheep and bullocks are starving for food and water. Hardly any are killed *that have not fainted first!*"

I thought that I should faint too; and I could only gasp despairingly, "But surely there is plenty of fish?" "Generally, ma'am, but not when the rollers are in." Utter collapse!

I hastened home sadly from my foraging expedition with a tale of want and woe; but so strongly did the

comic element prevail in the recital, that David and I
broke into peal after peal of laughter, and that was
almost as good as a meal. "Never mind the larder
just now," said he in his man-like way, "come and
see the croquet ground."

"Croquet ground!" I repeated, as a thought of
Nebuchadnezzar and his way of living crossed my
mind. "Can we eat grass?" But I might have
spared myself the question. Here was no soft inviting
turf for noiseless balls to glide over, no pretty green
carpet to deck with puzzling white hoops, no waving
boughs to shade the heated combatants and cool the
temper of the vanquished. Oh, no! The imagination must paint no such picture as this. The croquet
ground behind Commodore's Cottage meant a level
piece of glaring white concrete, about thirty yards
long and fifteen broad, with a close paling on the
further side, probably to keep off the dust and cinders,
while on the side nearest to the cottage a few withered
aloes, with tattered dust-stained leaves, struggled for
bare life. Such was the croquet lawn that I was led
in triumph to admire.

"Croquet here?" "No, of course not: but don't
you see it is the very place for the Observatory?
So level and stable for the piers; so near the house;
and as Cross Hill only cuts off 10 degrees of horizon,
it will not be in the way of Mars." "To be sure!"
I said; "just as if it had been made on purpose."
In our satisfaction at this happy discovery, we quite
forgot the bare state of the larder, which, I must not

forget to add, became strangely metamorphosed by the time we felt in want of a meal. I have never quite learned how, I only know that the effect produced on our larder by the thoughtful kindness of Ascension neighbours was the reverse of what happened to the cupboard of Old Mother Hubbard.

Somehow it all came right; and sitting that first evening after sunset in the verandah which looked upon our novel croquet lawn, we could speak of nothing, think of nothing, but the beauty of the heavens. Though Ascension was barren, desolate, formless, flowerless, yet with such a sky she could never be unlovely. The stars shone forth boldly, each like a living fire. Mars was yet behind Cross Hill, but Jupiter literally blazed in the intense blue sky now guiltless of cloud from horizon to zenith; and, thrown across in graceful splendour, the Milky Way seemed like a great streaming veil woven of golden threads and sparkling with gems. The Southern Cross—a poem in the heavens—shone out a bright welcome to us, while our old friend the Great Bear still kept faithful watch in the north over our wanderings. How strengthening and restful after fatigue and petty worry, is such an hour! One forgets to be careful and troubled about many things, and the soul trembles with its load of love and gratitude to Him who "made the stars also."

CHAPTER V.

ASCENSION PAST AND PRESENT.

Ascension discovered.—Its volcanic origin.—Its shape and size.—
"True as the needle to the Pole."—Embedded turtle.—A French
opinion of Turtle-soup.—The Sailor's Post-office.—Visit of Abbé
La Caille—British occupation of Ascension.—Why we keep it.—
Its peculiar government.—The Governor's troubles.—A decision
worthy of Solomon.—The population in 1877.—Ascension, the
Flora Tender.—Sea life ashore.—Ascension mutton.—A gallon of
water a day.—Novel domestic economy.

In order to make the process of taking root on
Ascension intelligible, it is necessary first to explain
something of the nature of its soil, and the peculiar
manner of its cultivation. In other words, to make
our own particular story less incoherent, it will be
advisable, in the first place, to tell the little I know
of the past history of our new home, and in what condition we found it in 1877.

Like its upturned face, the history of Ascension is
featureless and colourless, being only redeemed from
utter inanity by its contradictoriness. Doubtless there
were stirring times here once on a day when Vulcan's
forge was alight, but that was before we short-sighted
mortals dared to peep into this now deserted workshop
of the grimy god.

On Ascension Day, 1501, Juan de Nova, the great
Portuguese navigator, found the fire gone out, and
only hills of cinder and plains of ash to bear record
of past labours. Ascension, so called by its dis-
coverer from the fact of his having sighted it on
Ascension Day, is one of the peaks of a submarine
volcanic ridge which separates the northern and
southern basins of the Atlantic, and is situated in
lat. 8° S., long. 14° W., almost midway between the
coasts of Africa and South America. It is one of
the most isolated islands in the world, and has no
land nearer than St. Helena, which lies 800 miles to
the South-East.

Doubtless the apex of a great volcanic upheaval,
which the deep Atlantic could no longer hide, Ascension is now at rest. Not the slightest trace of volcanic
action has been recorded during the 377 years that
have elapsed since its discovery, but the absence of all
vegetation, and the slow progress of disintegration,
owing to the dryness of the climate, give it every
appearance of recent disturbance. Although there is
no record of recent disturbance on Ascension itself,
yet there have been observed at intervals, since the
middle of last century, certain volcanic phenomena in
its neighbourhood.

Both in the "Nautical Magazine" and in the
"Comptes Rendus" of 1838, accounts are given of a
series of marine disturbances in the open sea, between
longitudes 20° and 22° West, about half a degree south
of the equator. "These facts seem to show that an

island or an archipelago is in process of formation in the middle of the Atlantic; a line joining St. Helena and Ascension prolonged, intersects this slowly nascent focus of volcanic action." So perhaps Ascension may one day shake out her skirts suddenly, and frighten the greedy sea into giving her up a little more land.

Meantime our little island has an area of 38 square miles, and takes the form of an almost equilateral triangle, each side of which is about seven miles in length. The west side lies nearly north and south, its extreme angles being rounded off; but the eastern angle terminates in a well-defined point. Round the shore are black and rugged streams of basaltic lava, many of which can be traced to points of eruption at the base of Green Mountain—a great mass of trachyte 2,870 feet high, near the centre of the island—or to numerous little red-coloured hills that are scattered over the plains and northern and western borders. This reddish colour is owing chiefly to the large proportion of iron contained in the lava, as we discovered to our personal inconvenience.

On consulting a compass on one occasion, in order to determine our whereabouts, we were much surprised to see the needle point to what was, according to our preconceived ideas, the south! and still more surprised were we, when, on moving the compass some little distance, the fickle needle wheeled right about! Then we tried it at the former spot, and again the needle changed its pointing, so we removed

the loose surface to investigate, and found that a big
lump of red lava was the cause of this extraordinary
behaviour on the part of the compass. "True as the
needle to the pole" evidently does not mean much
when there is iron-stone in the neighbourhood.

Every specimen of rock that we were able to find
here was purely volcanic, except in some of the little
bays, where there are immense accumulations of small
water-rounded particles of shells and corals, inter-
mixed with a few volcanic particles. At the depth of
a few feet these are found cemented into a sandy limestone, of which the softer varieties are used for building. It is said that these particles become united in
the course of a single year, and I certainly found that
the turtles' eggs, deposited in the sand, get enclosed
in the solid rock before the sun has had time to hatch
them. A specimen of limestone in which this was
the case, was given to me by one of the marines
employed at South-west Bay to quarry stone for the
limekiln, and Sir Charles Lyell has shown a figure
(Principles of Geology, book iii. chap. 17) of some
eggs containing the bones of young turtles found thus
entombed.

From time to time, at long intervals, we have short
glimpses of the little island after the reign of Vulcan
and under the undisputed sway of the sea-birds.

One, Mons. Brazen, writing in 1726, tells us that
Ascension was discovered by Tristan d'Acunha in 1508;
but how could this be when Juan of Portugal discovered it seven years before? The old Frenchman

must have been dreaming, possibly after a too hearty
meal of turtle-soup, for he goes on to say of the turtle,
" On en peut manger tant qu'on veut, sans crainte
de s'incommoder." That depends on circumstances !

In 1673 Ascension was visited by the Dominican
Father Naverette, who speaks of it then as the " Sailor's
Post Office." " Mariners of all nations being accustomed at that time to leave letters here, sealed up in a
bottle, in a certain known cranny of some rock, to be
taken away by the first ship which passed in an opposite direction."

We hear that a man of mark lighted on the island
in the month of April, 1754. At that time Abbé La
Caille spent five days there ; but evidently without
seeing much that he considered worthy of record, for
all he tells us is, " Ascension est une espèce de butte
en pleine mer. Elle est couverte d'une terre rouge,
semblable à de la brique pelée. Sa capacité est un
gouffre, qui retentit d'un bruit sourd et confus lorsqu'on
frappe le sol des bords du volcano."

Portuguese and French alike passed the untempting
isle. No nation coveted its barren shores, until the
British lion stretched out a paw in 1815 and gathered
it into his heap of treasures. Napoleon had then
been sent to St. Helena, and we dared not leave such
a vantage point open to the enemy ; so the British
flag was planted on yet another spot of the globe, and
Ascension became, to all intents and purposes, a man-
of-war guarding Napoleon at St. Helena. Though
there is now no Napoleon to guard, we still keep pos-

session of Ascension, for no other reason, that I can
see, than that we do not wish anybody else to have it.

True, it is useful as a coaling station, and the fresh
trade-wind, constantly blowing across its flowerless,
waterless plains, brings health to many of our poor
sailors who have drunk in the blood-poison of the
Gold Coast swamps. But need we spend £40,000 a
year for this, when St. Helena might make a better
Sanatorium at half the expense? There the S.E.
trade-wind blows still more freshly, and cool showers
fall to beautify everything, and to supply water and
fruit and vegetables in plenty for the sick.

In the Ascension hospitals—of which there are two,
with a staff of three medical officers—the want of these
advantages is much felt; and moreover, the cheerless,
changeless surroundings are likely to have a depressing influence on nerves already weakened by fever and
ague.

All this must tell against Ascension as a naval
hospital; but probably the Admiralty may find some
advantage in having the place perfectly under their
own control, and thus being able to keep the men in as
perfect discipline as when afloat. In case of war, also,
Ascension, in the hands of an enemy, might be the
means of inflicting serious injury on our commerce—
lying as it does, directly in the track of our merchant
ships on their way to and from the East. So we keep
as a friend what might prove a dangerous foe, and pay
dearly for our white elephant, rather than allow anybody else the privileges and expenses of possession.

The government of Ascension is unique. No other *land* in the world is ruled by the same laws, and my husband and I are the only civilians that have ever been subject to them. When David decided on this island as the most favourable spot on which to observe the Opposition of Mars, the first step was to obtain permission from the Lords of the Admiralty to go there. This permission was readily granted, through the kind intervention of the Astronomer Royal; and not only that, but, what was of immense importance to me, the accommodation usually accorded to a married officer was provided for us. Our official letter also contained the promise of assistance in erecting the Observatory, a blue-jacket for night-watch, and a gracious permission to *buy meat*. Without this letter, we could no more have landed on Ascension than we could have boarded a line-of-battle ship.

The sway of the Captain is quite as absolute in the one case as in the other, and, generally speaking, the same regulations apply to both. But there is one notable exception. A certain number of women and children are allowed on board this " ship ashore," which of course has the effect of somewhat relaxing the discipline. Indeed, it is on record that a certain Captain was so perplexed by his difficulties in governing the female portion of his crew, that at last he gave up the attempt in despair.

But à propos of these difficulties of his, there is a story told, which, if true, shows that on one occasion at least, he proved himself no mean diplomatist.

Once upon a time, a fierce war raged between two dames
of, let us say, equal degree. The apple of discord
was—precedence in church ! Each claimed her right
to the pew in front of the other, and both being
equally determined, there seemed no way of settling
the question except by referring it to the chief. This
in due course was done, and the decision is worthy of
Solomon: "Let age take the higher place," said the
wily Captain, "and let the younger lady give way to
the elder ! " From this there was no appeal, and next
Sunday, lo ! both ladies were seated in the pew nearest
to the door. The lower place was now the prize, but
whether a new war, more desperate than the old,
broke forth in consequence, and whether the poor
Captain's troubles turned his brain in the end, neither
history nor tradition sayeth.

This is a legend of the island, but I have more
respect for my sex than to believe it, and the state of
society on Ascension in 1877 strengthens my unbelief.

At this time I found myself the sixth lady on board;
and a few of the men are allowed to have their wives
with them, on condition that the latter make themselves
helpful to the community in some way. The male
population is under 200, and consists of a company of
marines, a few blue-jackets, several St. Helena boys
who act as servants to the officers, and 70 or 80
Kroomen—a fine race of negroes from Kroo country on
the West Coast of Africa, about whom I shall have
more to say presently. All these men are of course
under the strictest naval discipline, and the Captain's

office is the "quarter-deck," where every offender, from
the greatest to the least, is judged and sentenced.
Everything is the property of the Government, and
each officer and man receives his daily ration of bread,
meat, and "grog," just as on board ship.

Indeed, in the *Naval Gazette*, the population of Ascension will be found under the heading "Crew of the
Flora Tender;" and service here does not mean half-pay
to the naval officer, but counts for active service afloat.
Ascension acquired the name of the "*Flora* Tender,"
I believe, at the time that H.M.S. *Flora* was anchored
there, and when the island of course provided her supplies. Now the *Flora* is stationed at the Cape for
better anchorage, but her "Tender" still stands firm
in mid Atlantic, and never drags her anchors as the
Flora once did alongside of her.

It was late in life for us to go to sea, but we very
soon dropped into sailor-like ways, and by-and-by we
adopted even the language of Jack. A kitchen was
not a kitchen here, but a "galley;" the pantry became
a "locker;" our floors and tables were no longer
scrubbed, but "swabbed out;" and the dinner had
not to be cooked but to be got "under-weigh."

The only regulation we mutinied against was,
"Lights out at 10 P.M.;" and for this rebellion we got
a free pardon, no doubt on the ground that an astronomer, being a species of lunatic, is not amenable to the
laws of any country.

We were indeed in a new country, and one that
taught us, with many other things, the fallacy of the

belief that " Gold commands everything." Not even a
Rothschild could buy a juicy leg of mutton here, nor
enjoy the luxury of a fresh salad with his cheese.
That mutton ! Shall I ever forget it ? Our first
" gigot," of hock-bottle shape, would have made an
English butcher faint, and ought to have been sent to
the British Museum, there to consort with time-tough-
ened mummies, and testify to future generations the high
state of training attained by Ascension sheep in 1877.

Poor sheep ! They were almost starving ; and so
were the miserable, gaunt-looking bullocks, that we
sometimes saw prowling around the house at night,
having wandered over five miles of terrible road from
Green Mountain in search of food and water. I could
not bear to see them in such distress, and yet we could
not relieve them, being ourselves limited to *one gallon*
of water a day for all purposes, and our whole allowance would have been but a drop on the tongues of
these poor animals.

This scarcity of water it was at first very difficult to
take into account in household expenditure ; and my
surprise was great when, on the first morning I sent
some linen to be washed, " Sam," our handsome Krooman, returned to say that I had forgotten to send the
water. This was truly an extra thought to the housewife ; and in many ways the first days of housekeeping
on Ascension were rather bewildering. But by-and-by
light appeared through the wood, and I found that
once started on the proper routine, the road was not so
rough after all.

By careful management and a plentiful use of salt-water whenever it was practicable, we could eke out our scant allowance of fresh water to a sufficiency; and this novel poverty enabled me to make two valuable discoveries in culinary art, viz., that fish and potatoes are better when boiled in salt water than in fresh. We soon got accustomed to tinned milk and vegetables; and when the rollers disappeared, we found ourselves by no means dependent on the scanty meat rations, for the fish here was as good and plentiful as it had been at St. Helena. And then there was the turtle!

Surely Ascension should be the paradise of Alder-men. The first spoonful of that clear, creamy nectar called Turtle Soup, is enough to reconcile any gourmet to banishment here for life! A turtle was killed once a week, and our share of the booty generally pro-vided us with sufficient to make a turtle-steak pie, besides a slice of fin for soup. The steaks were excel-lent, stewed or baked, but they could not stand the ordeal of a gridiron. Cooked over the fire, the meat became hard and juiceless, almost as bad as an Ascen-sion *beef* steak. With the fin, and taking care not to omit the "calipash," and "calipee" we made delicious soup, when we could spare water for it; but some weeks we had to pay the price of a little extra extrava-gance in the precious fluid, by being deprived of our soup. Then, with sad hearts, we stewed the fin, and it made a palatable if not a pretty dish.

Verily, all one's pre-conceived ideas of the relative values of things were here turned upside down.

Water carefully measured and treasured, potatoes 4*d.* per lb., occasional cabbages from St. Helena knocked down by auction at 1*s.* 6*d.* each, milk priceless, and turtle soup for nothing. It was very difficult to comprehend at first, and I suffered much from alternate feelings of stinginess and prodigality before being able to master this new domestic economy; but after the first feelings of bewilderment were over, the novelty was delightful. Something like the fresh, stirring sensation of a shower-bath, after the head has recovered from the first shock of the falling water.

CHAPTER VI.

ROUND ABOUT GARRISON.

OUR first work was, of course, the Observatory, for observations ought to begin on the 17th of July, and it was necessary that no time should be lost in getting ready.

Our twenty tons of baggage had been landed on the evening of our arrival. At 6 o'clock on the following morning, carts were busy bringing up the numerous cases from the pierhead, and marines were at work unscrewing box-lids and unsoldering tin-lined cases. The sound of hammer and saw woke an echo in the still morning, and by breakfast time the croquet ground was littered with extraordinary heaps of queer-shaped materials.

In the south-west corner, masons were laying down a level bed of cement for the sleepers of the circular railway on which the Heliometer House revolves; for, as stars had to be observed in all parts of the heavens, the opening of the Observatory must neces-

sarily be arranged to view any part of the sky.
This was managed as follows.

A strong octagonal frame was mounted on flanged
wheels rolling on the railway, and this frame carried
a structure of iron gas-pipes, screwed together and
stiffened by cross ties of iron wire—the whole forming
a species of cage. To make this a good water-tight
house, it was necessary then to cover it with canvas,
previously shaped and fitted with means of attachment
to the frame. My husband was no stranger to the
details of this portable Observatory, which he had
already used in Lord Lindsay's Transit of Venus
Expedition at Mauritius; and the naval artificers of
Ascension proved willing and intelligent workmen.

Before night the iron cage was revolving sweetly on
its well-oiled wheels, and a couple of blue-jackets were
lashing on the canvas cover. The process of fixing
the canvas was very like that of reefing a sail, so Jack
was quite at home; and then the rolling dome was so
like the revolving turret of an ironclad, that in this
novel combination of sea-machine and sea-work
ashore the sailors took great delight.

In the north-east corner, some marine carpenters
were busy putting up the Transit Hut. This Observatory has also a history. It had been constructed
at Greenwich for the British Transit of Venus Expedition in 1874. It had been in Egypt, mounted on the
Mokattam Heights above Cairo ; and probably many a
Pacha had smoked his cigarette under its cover when
visiting Captain Orde Brown at the Venus station there.

Like everything emanating from our National Observatory, it bore the stamp of the order and method which characterize our Astronomer Royal. Every screw and every plank of it were so marked and numbered, that, almost without instruction, the marine carpenters could fit it together, after David had laid down the lines of its position.

This Observatory, unlike the Heliometer House, was fixed, and its shutter opened due north and south, so that the Transit Instrument inside might be capable of being directed to any celestial object as it crossed the meridian.

It was indeed a busy day for my husband, for morning and evening " sights " had to be taken with the reflecting-circle to determine the error and rates of his chronometers. The shadow of a plumb line had to be laid at noon, to give the line of direction for the foundation of the Transit Hut; and all day long he toiled in shirt-sleeves, unpacking the more delicate parts of the instruments, and trying to be everywhere at once, so as to keep all going.

That evening there was, to all appearance, a complete Observatory on the croquet lawn; but much had still to be done. The Transit Hut was outwardly finished, but the piers inside had yet to be built, and the instrument mounted thereon and adjusted exactly in the meridian. With regard to the Heliometer Observatory, though the heavier parts of the instrument were *in situ*, the tube and more delicate parts remained to be attached, and the whole had to be afterwards adjusted accurately by observations of stars.

However, the third day of steady hard work, by night and by day, saw all in order, and an Observatory established, as complete as the heart of travelling astronomer could desire. Our Observatory staff was also completed by the addition of an intelligent blue-jacket, by name " Graydon," installed as night-watch and lamp-bearer.

The anxiety of Observatory building being off David's mind, and the chief difficulties of the commissariat department off mine, we began to look outside ourselves a little, and, after the fatigue of so much change and so many new sensations, to enjoy the rest of one day repeating itself in the next.

We began to think of little excursions when the sun drew near the west horizon, and our first walk had for its object a visit to the famous Turtle Ponds. These lie close to the sea, at the north end of Garrison, just where Long Beach terminates in rock under the fort; and are simply two large stone-built enclosures, into which the sea flows freely through narrow sluices. Here I saw more than a hundred huge creatures, looking like monsters of a bygone age. At first sight these dark masses, just showing above water, might be mistaken for slimy, seaweed-covered rocks, till one of them slowly moves—places a finny foot on the top of the " black thing " next to it, and rears aloft an ungainly head, showing a breast of leathery, shrivelled skin, speckled and streaked with a motley of yellow, green,

brown and red. The small head and the slow, undula-
tory movement of the neck, betray a member of the
reptile family, but here the serpentine character ends.
Everybody is familiar with the shape of what Wordsworth calls—

> "A shell of ample size, and light
> As the pearly car of Amphitrite,
> That sportive dolphins drew."

Only Wordsworth, I think, could have found anything so pretty to say about the turtle shell, which
entirely encases the unshapely creature. Those we
saw were certainly of ample size, each animal weighing
from five to six hundred weight; but they take a long
time to acquire this weight, and the full-grown ones
are said to be a hundred years old.

I do not know how this conclusion is arrived at, for
the young turtle are seldom, if ever, seen from the
time that they make their way into the water, straight
from the egg, until they return again to land, at full
growth and maturity, to deposit their eggs; but they
are certainly slow-moving, slow-living, slow-growing
animals.

The marine cooper in Garrison had, with great
difficulty, reared a turtle from the egg until it was six
months old. At that age it could easily be covered by
the hand, and as the baby turtle, newly hatched, is
about the size of a Dorking chick, at this rate of
growth it must needs take a long time to develope into
a gigantic animal of six hundred weight. Some have
been known to weigh as much as eight hundred weight,

for this species of turtle (*Chelone Mydas*) is quite a giant compared to the little turtle of the West Indies. The latter, however, is the more delicate for table use, and is the favourite in the London market.

The green turtle of Ascension, being not only less delicate in flavour but more delicate in constitution, is very difficult to convey to England alive, especially in winter. Many are the pathetic stories told of poor doomed turtles, lying on their backs on ship-board and sobbing their lives away, thereby causing the expectant recipient to turn homeward from the London Docks a sorrowing and soup - disappointed man.

Captain Brandreth, R.E., who visited the island officially some forty years ago, says, — "In 1830, the commandant freighted the transport with sixty of the finest flappers that the season had produced. They were destined for some of the most distinguished individuals in England, and the largest and finest was for His Majesty, with instructions, if but one survived, it should be considered as so appropriated; the commandant acting as nearly as possible upon the principle that the king never dies. And the precaution was by no means unnecessary, as, in fact, only one did survive. To prevent intrigues in favour of particular patrons or friends, each turtle was marked, on his fair white belly-shell, with the name of the owner, and the sailor in charge of the party duly reported each morning their state and condition, as thus :—' Please, your honour, the Duke of Wellington died last night,' or, ' I don't like the looks of Lord Melville this morning, sir.' Then followed certain interesting questions :—' How is the Lord Chancellor?' ' Why he looks pretty lively, sir,' and so forth."

One of the many curious facts connected with the turtle is, that no males are ever seen. The females are captured when they come to lay their eggs on the little sandy beaches that run here and there into the rocky coast of Ascension. At North-east Bay, South-west Bay, Dead Man's Beach, &c., there are men stationed during the turtle season, from Christmas to Midsummer, to watch for the unwary turtle as she scrambles up, about a hundred yards above high-water mark, to deposit her eggs. Here she digs three or four nests for herself, one after the other, eight or ten feet across by about two feet deep. In these she lays often three hundred eggs in a season; forty or fifty in each; and leaving them to incubate in the hot sand, a two months' process, she makes for the water again.

The men are forbidden to capture the turtles until *after* they have deposited their eggs, but as the cautious mothers often perform an unsatisfactory manœuvre called " making a horse-shoe " (that is, they come ashore, and not finding a place to their liking, simply describe a semicircle on the beach, and return to the water without making a nest at all), the men are eager to seize the poor things as soon as they appear, for the sake of the half-crown awarded for each turtle.

For turtle-hunting it is necessary to be armed with a stout stick and a noose of rope. The noose has to be slipped over a back and fore fin, which, by this means, are drawn together, and the rope is wound up on the stick till it touches the turtle's upper shell; this forms a lever by which the creature is turned over. Once on her back the unhappy turtle is perfectly help-less, and in this way an average of 300 turtles are now collected every year from the various breeding-places, and transferred to the ponds on Long Beach, there to wait the evil day of soup-making.

Formerly the number was much greater, so many as 2,500 turtles having been turned in one year; and it is said that in the " good old times " of Ascension any ship's crew landing here might have turned fifty in a night. Connoisseurs say that the turtles fresh from the sea are better fare than those kept long in reserve; so if any remain when the new stock arrives, they are restored to liberty. But, by the less fastidious, turtle is eaten with equal relish after the animal has lain in the ponds for a year without diet of any kind. They live on " nothing a-year "—happy turtles! Of course wise men tell you that they feed upon sea-weed, and the crustaceans and molluscs which are washed into the ponds with the salt-water, but surely these can be to them but the veriest bon-bons, and I prefer to believe that they dine but once a-year! Then I can also believe in their great age, and look upon the finny

monsters with awe, as the last survivors of the ante-
diluvian age, when life was lived slowly, and ninety
years were but as the childhood of a man.

It was so hot during the day, that we, in common
with our neighbours, preferred to shut ourselves up
within jalousies as much as possible. Thus it happened that the first time I peeped abroad in full sunlight was to go to church, and the tinkle, tinkle of the
little bell gave me strongly the feeling of home. There
was something exquisitely soothing and comforting too
in the quiet worship of God in this isolated tabernacle,
surrounded as we were by the bowed heads of so many
of our brave sailors.

St. Mary's, at Ascension, shares, with the other
buildings in Garrison, the monotonous level that lies
between the foot of Cross Hill and the roadstead.
There is little of the ecclesiastical in its exterior,
except it be the primitive belfry, containing a single
unmelodious bell which is rung in rather a primitive
way by pulling the clapper. The outside walls are of an
ochre yellow, flecked with green jalousies, which shade
the glassless windows. Through these the cooling
breeze steals in and just stirs the leaves of the open
prayer-books, then, with a hushed whisper, softly es-
capes, as if afraid of having touched too rudely some
holy thing.

Within, near the door, stands a tiny baptismal font
of the soft grey limestone of Ascension, and as you
pass along the aisle, you note the many loving tributes
to the memory of dead comrades, which, with their

simple inscriptions, give a pathetic interest to the plain rough walls.

As to ornament the little church has not much to boast of, except it be that of many low deal pews filled to overflowing with earnest-faced men, each one the very picture of cleanliness and order in his fresh Sunday jacket, innocent of stain or crease. What a splendid dress is that wide-throated blue jacket, and how well it suits these frank fearless faces, where the bright, intelligent eyes tell that training and discipline have not made the machine and unmade the man!

Hardly a treble note softens the rough, hearty voices which fill the church with well-declared praise. David, in his civilian coat, is quite as exceptional as the Chaplain in his surplice; so that it was not difficult to imagine one's-self on board some ship at sea, and I almost expected to hear the sound of waves dashing against the outside walls.

Not only in outward seeming but in inward desire did we differ somewhat from our neighbours here, for now a great longing for rain was abroad—a longing not unnatural after a drought of fourteen months, with only one poor condenser on board, capable of providing no more than twelve tons of water per week for nearly two hundred souls and many thirsty brutes. This seemed improvident on the part of the powers that be, and our Chaplain so resented it on behalf of his flock, that he flatly refused to pray for rain, saying it was England's duty to supply the " *Flora* Tender" as well as the other ships of her navy, with proper condensers.

We shared this opinion, having come many miles to escape from clouds and rain ; and when the weather conversation was afloat, we alone were silent and rejoicing.   But alas! before very long it seemed as if our rejoicing was to be turned into mourning, for not only clouds, but rain came to damp our selfish joy. During the nights of the 25th and 26th July some showers fell in Garrison from the heavy masses of cloud that had for many nights previous, rolled over Cross Hill from Green Mountain, sadly interfering with astronomical work.

On the 17th the Observatory had been ready for duty ; but no sooner were the instruments adjusted and some preliminary observations made, than the face of the heavens darkened and we began to fear.   For five nights no work was done, and on many nights following only interrupted observations could be snatched from between the clouds.   It was a treacherous sky, and I wasted much time in watching it.   After shining upon us with unremitting fury for twelve burning hours, the sun would set over the sea in a wealth of flame, leaving the cloudless heavens flushed with a proud memory of his departed glory.

This pink after-glow is quickly succeeded by the sudden night of the " gloamingless " tropics, and as the blue-black vault o'erspanning us begins to sparkle with lesser suns, we long impatiently for Mars to rise over the crest of Cross Hill.   But alas! from the

birth-place of Mars a tiny white cloud arises no bigger
than a man's hand. It looks soft and harmless enough
at first, but while you watch, the snow-white mass gets
streaked with grey as it lengthens across the sky, like
some serpent monster gathering strength and darkness
in its course.

And Mars ? We know he is behind that envious
cloud, and we watch for a rift. It comes: the sky
seems cleansed by the trail of the serpent, so pure is
it ; and in the twinkling of an eye the telescope is
turned on the planet. But to no purpose, for again
a fleecy cloudlet peeps over Cross Hill, and, rushing
swiftly onwards towards Mars, soon wraps him in a
soft grey mantle, and again he is lost to us.

These " rifts in the clouds " were for me moments
of intense excitement, and, knowing how many minutes
were required for each measure, I watched sky and
chronometer with aching eyes, dreading lest the advancing cloud should give too little time for any work,
as it chased its predecessor across the sky. They
were very beautiful too, these streaks of blue —
so bright and pure—with Mars brighter and purer in
their midst—like some noble river rolling between
snowclad mountains and wearing a diamond on her
bosom.

Yet this lovely, changing sky I could not love, for
empty pages, where figures should have been, lay open
by the Heliometer ; and my husband looked weary,
while Graydon and I yawned in concert. Many hours,
indeed whole nights, this went on, and sometimes the

clouds followed each other so rapidly that no measures
could be secured at all.   Then I was seized with an
insane desire to get beyond this nest of clouds.   But
it is not so easy to pick up a Heliometer and walk over
a hill with it ; and there really seemed nothing to be
done, but to fold our hands in idleness and wait for
the silver lining.

CHAPTER VII.

A NIGHT ON THE CLINKER.

Thus passed two weary weeks. We pored over dry
statistics, hunted up every scrap of weather record, and
annoyed everybody with questions about cloud and
wind; but to little purpose.

The crew of the *Ascension* is a changing one, three
years being the usual term of service, so that no
one was able to give us the benefit of long experience
of Ascension weather. The answers to our questions
were contradictory and distracting in the extreme,
being based on casual observation or general impression. The only thing that everybody seemed to agree
about was, that " such cloudy weather had not been
known within the memory of the oldest inhabitant; it
was altogether exceptional, and by-and-by there would
be as clear skies as even *we* could desire." Yes, "by-
and-by" perhaps; but meantime Mars would not wait,

and the present was threatening our expedition with
disaster.

Oh! those weary weeks. Fearful of losing a single
hour of star-light during the night, we watched alternately for moments of break in the cloud, sometimes
with partial success, but more frequently with no result
but utter disappointment, and the mental and physical
strain, increasing every night, grew almost beyond our
strength. What was to be done? There was the Observatory complete, the instruments faultless, and the
astronomer idle, for there too was the cloud. Sometimes it "curtained the sky from pole to pole," but
more frequently it confined itself to the narrow, snake-
like band, stretching across the zenith and showing
clear sky to north and south.

Is anything so temper-trying as waiting—anything
so heartbreaking as enforced idleness in the midst of
work? I bore it badly, or rather I didn't bear it at
all, and the peevishness of disappointment was fast
giving way to a sullen despair in my mind, when one
day David spoke and took away my breath. He said,
"Let us prove how far this cloud extends, and find
out whether there is any accessible part of the island
not covered by it."

This proposal from him startled me, and the idea which
had so often floated idly through my brain, now for the
first time assumed a practical shape. It gave speech
to silent desires which I had been ashamed to utter.

I could not calculate the risks, and therefore I was
bold; but now they were well considered, and the re-

sult of sober thought agreed with my visionary longings. " Let us explore, and move if need be." It was like the order to attack after the hateful inaction of a siege, and I was eager to be up and doing.

It was now the 30th of July. The sun had set. The after-glow had faded away, and stars were shining everywhere but just where they ought to shine. Green Mountain was, as usual, busy dispatching a train of cloud over Cross Hill, and right in the path of Mars. It was a typical night, and we determined to act; but how ? David could not leave his post in Garrison, lest an opportunity for observations should occur, and, after the fatigue of a hot day's work, we could not ask any of our neighbours to give up the rest of the cool night for us. So I offered myself as pioneer, but my offer was at first rejected with some decision. "Impossible ! You have never been beyond Garrison; there is no road; there may be dangerous gullies; and wild cats infest the plains—you would find the walking bad enough by day, and at night *impossible*." But having a considerable leaven of Luckie Mucklebackit's spirit in me, I meant to try hard for my own way; and after showing how carefully I had been studying the Admiralty chart—studying it till every crater and every watercourse seemed stamped upon my brain, the "impossibles" grew fainter till finally they disappeared. The chart was brought out, *not* for the first time, and I was proud to show off my knowledge.

For some little way south of Garrison a winding

road, or track, was indicated; then open plains of
volcanic ashes; while beyond, and everywhere bordering
the sea, rose up great lava rocks, or, as they are locally
called, " clinker." Now the question seemed to be,
does the cloud extend beyond the plain, or does it
not? If it does, the case is hopeless, the general
character of the " clinker " being, that it is inaccessible to all but goats and adventurous donkeys.

It was 9 P.M., and no time was to be lost. Hill was
called upon for his knowledge of the way, and we found
it *nil,* but he thought that Corporal B—— knew something of it.

" Then," said my husband, " ask Corporal B——
whether he will accompany Mrs. Gill in a walk towards
South Point at 10 o'clock, and you will go too with rugs
and a luncheon basket." Hill looked rather mystified,
as well he might; but the order was one after his
own heart, and the old corporal showing a spirit not
less adventurous, my guides were ready and waiting
for me at the time appointed. My watch was carefully
timed with the chronometers, and David and I arranged
to make simultaneous observations of the clouds every
half-hour till 3 A.M., when I promised to return.

At 10 P.M. we started. My spirits were higher than
they had been for many nights, but David looked
anxious, and warned me again and again not to run
into any danger. I reassured him with boasts of my
knowledge of the chart and of the places of the stars.
Besides, in an hour the moon ought to rise, and in
the tropics she is a brilliant lamp; but in case of

the clouds thickening, we provided ourselves with a
lanthorn and a bull's-eye, to serve in her stead.

For the first mile I found the road very tiring—soft
and yielding, and bestrewn with loose lumps of clinker;
moreover I had made the mistake of putting on low
shoes. I chose them because they were thicker and
stronger than any boots I had, not considering that
the sand, or crushed cinder rather, would get inside
and chafe my feet. Our next misfortune was the
sudden rising of the wind, followed by the total eclipse
of our lanthorn. But, happily, we had our bull's-eye to
fall back upon, and by this light " dimly burning" we
proceeded for the first half-hour.

Now I observed that the cloud, instead of being
entirely overhead, just reached our zenith and then
dipped northward. This greatly encouraged me, and
as my eye got more accustomed to the darkness, I was
better able to choose the best of the bad way. It
was still very dark—the wind rose higher—the moon
gave no notice of her coming, and the weird ghostli-
ness of the little bit of surrounding that fell under
the light of the bull's-eye, I shall never forget. A sort
of awe, not unpleasant, but the reverse, was stealing
over me, and I felt just in the mood for an adventure;
when, lo! close to my ear, a shrill uncanny shriek rang
out through the stillness. The corporal flashed the
bull's-eye in the direction of the sound and disclosed
a dim, moving object, undistinguishable in shape or
colour. Then I thought of cats—of bullocks turned
carnivorous in their hunger, and my heart grew cold.

"It's Jimmy Ducks," said the corporal, in a tone of recognition.

"Who is Jimmy Ducks, and what is the matter with him ?" I asked, trying not to shiver.

"Oh, ma'am, Jimmy Ducks is an old mule, blind of an eye. He has been turned out on the clinker to pick up his living, and he is frightened at our light."

Oh dear! how very small I felt! Don Quixote and his windmill were not more absurd than Jimmy Ducks and my silly self; and so disgusted did I feel with my cowardice, that I almost forgot to note the cloud.

For some little way the road had been tending more to east than south, so that we were not advancing much, but only getting nearer to Green Mountain, over which the moon was now struggling with a heavy cloud; not succeeding in piercing it, but just throwing out its awful shades. I never saw such a cloud, and ceased to wonder at the width of its skirts. But now we were not *covered* by it as at Garrison; our east horizon was indeed entirely obscured, as it had been during the previous half hour, but the south expanse of sky was larger, and I felt all the excitement of running a race with my enemy.

There was no longer any road, and walking was easier. This may sound contradictory, but we certainly got along the hard weather-beaten plain at a much greater speed than we had done on the heavy road, loosened by traffic to South-west Bay, whence lime and turtle are occasionally brought by cart to Garrison. Because of the road winding so much to eastward, we had lost sight of the

sea some time before getting upon the plain, and now
that no path marked the way, the bull's-eye, borne in
advance by the corporal, began to waver, and so did
my confidence.   We were losing our bearings, and I
thought it wise to turn at once due west to catch sight
of the sea again.   Then we followed the line of coast
southward till, suddenly, the clinker rose up and
stopped the way.

It was now midnight.   Mars had about 30 deg. altitude, and just skirted the cloud which covered the
north and north-east, leaving the other parts of the
sky brilliantly bright.   Near him shone Saturn, a
glorious contrast in colour, and Jupiter blazed over
head as I spread my rug on the clinker, and tried by
looking hard to make "the darkness visible."   The
moon was still struggling with the cloud and gave out
a fitful light, just sufficient to show the utter barrenness and desolation around me.   Here and there to
eastward, small craters were tossed over the plain.
It was too dark to distinguish their colour or form
perfectly, but they all appeared dingy, and of a
uniform conical shape.   Behind us lay an uninter-
rupted flat called "Waterloo Plains," and in front
rough, needle-like masses of clinker pointed their
spires skyward.   These continued southward as far as
I could see, and down to the shore as well, for I
walked, or rather scrambled, in that direction, to try
whether I could find any path leading farther south;
and I convinced myself that *there*, there was no break
in the rocky wall.   To eastward I thought there might

be ; but as the night was still dark, I feared to miss my
way and make my husband anxious by being late. So
I sat down on the clinker and had a glass of water and
a biscuit, while my guides retired for their refreshment
and a smoke.

I wonder what they talked of, and if they thought
of me as a mad woman chasing a cloud by moonlight !
Doubtless they did. But absurd and aimless as my
chase appeared, the object I pursued was a glorious
one ; and, during the hour I sat on the Ascension plain
and watched the clouds, I felt as if it were within my
grasp. That is to say, I became convinced that this
cloud was partial, that it formed in the east over
Green Mountain and took a direction almost due west
towards the sea. I had now succeeded in getting
alongside of it, and I was nearly certain that all would
yet be well if there was any accessible way farther
south.

Before starting on our return journey, I got the men
to make a cairn of clinker, which we topped by a
bottle with a bit of red cotton tied to it, so that David
might see where we had been stopped, should he
decide to come and explore for himself.

The homeward road seemed very long, now that the
excitement of inquiry was over ; and though I did not
feel tired, my feet ached with the sand and small stones
that kept getting into my shoes. I was truly glad
when, after five hours' absence, I met my husband
looking out for me about half a mile from Garrison.
He seemed relieved that I had returned in safety ; but

congratulations were cut short by his anxiety for my
report, and I was equally anxious for his, for I could
not *really* know whether my expedition had done any
good till we had compared notes.

This we did the moment we got inside Commodore's
Cottage, and with the result of convincing David that
the cloud was systematic. This determined him to
move the Observatory to the extreme south of the
island at all hazards, providing Captain Phillimore's
consent could be obtained.

I was almost frightened at his decision, and, coward-
like, looked back after having put my hand to the plough.
For the first time difficulties presented themselves to
my mind in such legion that I could scarcely think.
I remembered how we had been told at first that Garrison was the only habitable part of the island, except
the cloud-capped mountain. I remembered the ruts
and the rocks, and thought of the Heliometer. I remembered the want of water, the absence of all local
habitation under a tropical sun ; and had it not been
for very shame, *I* should now have cried out " Impossible." But I kept silence and looked on.

My husband had already spoken to Captain Phillimore
about the advisability of some change, and the morning
after my excursion, or rather the same morning, at
7 A.M. they set out in the strong cart to look for my
cairn, and to try and penetrate south of it.

In some hours they returned with the news that
my landmark had been discovered about four miles
south, and that it showed our " track " to have been,

as I imagined, too close to the shore. More to eastward the plain ran farther south, and they were able
to drive between two little hills (Saddle Crater and
Round Hill), both of which, it seems, I had kept to
landward. At this point the clinker showed ominously
ahead, so they tied the horse to a big stone and
climbed the shoulder of Gannet Hill, now immediately
on their left, in order to reconnoitre. Nothing but
clinker to be seen; and that too of such formidable
character as to form an apparently insurmountable
barrier to the passage of the Heliometer.

An iron-bound coast indeed ; and on the south it
stretched into a narrow promontory (South Point)
bristling with clinker, erect and sharp as the quills on a
porcupine's back. On the lee (or north-west) shore
of South Point, there nestled a little bay ; and running
into it from the plain was an empty watercourse, which
seemed, with its more rounded stones, to afford some
possibility of reaching the sea. A rough and rugged
road it proved, however, when Captain Phillimore and
my husband tried it ; and certainly not available for
the transport of heavy and delicate instruments. Oh !
for an hour of Prince Houssain's magic carpet, to
bear the Observatory through the air ! for here is the
very place for it, high upon the rocks above this little
bay. Green Mountain and its clouds well to northward, nothing but this strip of barren rock to the
south, and Mars can be seen to set, and almost to rise,
over the sea. Here there can be no systematic cloud,
only what the trade-wind brings, and we could fare

no better even by getting into the sea. It was very
tempting, and the explorers looked at land and sea,
and at sea and land. Both routes seemed dangerous
—the land route indeed impossible, while the surf and
rollers which beset the Ascension coast gave little
hope of the sea. No landing had ever been made
at this bay, but as there was a tiny bit of sandy beach,
Captain Phillimore thought it might be attempted,
should my husband make up his mind to run the
risk.

Such was the exciting news brought home to me;
and the multitude of questions, fears, and anxieties it
stirred up in my brain made me feel quite giddy, and
very thankful that I was not called upon to decide.

Oh! the sickening responsibility of making up one's
mind, and choosing between two evils. I had no
word to say, and could only share the weight of
anxiety without being able to lighten it. Either way
looked gloomy. On the one hand, my husband felt—
" If I stay here and fail, I shall have failed also in my
duty,. not having done my utmost. On the other
hand, every night is now of importance, and a week
is lost *certainly* if I pull down the Observatory, while
the slightest accident to an instrument here, with no
one to repair it, will be fatal to the expedition."

Yes! both " ifs " were unpleasant, but the first was
intolerable, and after a day of anxious thought, David
made up his mind that an attempt to reach South
Point must be made. After coming to this decision,
our great aim of course was to carry it out with as little

delay as possible, and to this end Captain Phillimore
kindly promised every assistance in his power. Water
should be condensed for us at once ; tents, lime, bricks,
cement, coals, cooking-stove, &c., looked out from the
naval stores, and a party of marines should accompany
the goods to South Point and see everything in order
before returning to Garrison.

Without knowing the limited resources of the
island, and the amount of regular work to be gone
through every day, one cannot realise the exceeding
energy and good will embodied in this offer. I always
feel a lump rise in my throat when I look back upon
this time, and remember how difficulties were removed
from our path, not only by Captain Phillimore, but by
every officer under him, as well as by the officers
of the ships then in harbour. Indeed, our kind
friend Captain Hammick of the *Cygnet* levelled the
last mountain by offering to send his Kroomen to carry
the Heliometer-tube and the more delicate instruments
over the clinker by hand.

It was on the 31st of July that the important decision was made, and, strange to say, that same night in
Garrison, my husband was able to make his first complete determination of the parallax of Mars. The sky
was cloudless from sun-set to sun-rise, and I wavered,
wondering, as many others did, whether the new hope
would shake the new decision ; but when I questioned,
the answer was, " The man that hesitates is lost."

CHAPTER VIII.

CHANGE AND CHECK.

Rooted up.—A strange procession.—"Sister Anne! Sister Anne! do you see anybody coming?"—Good news.—Mars Bay.—The Heliometer and the Kroomen.—All's well that ends well.—A solitary palm.—My first sight of Mars Bay.—Clinker flooring.—Roller fever.—At work again.—Sam and Fetish.—My second coming to Mars Bay.

At daybreak on the 1st of August, David was hard at work with the men, dismantling the snug little Observatory. Again the sound of tools was heard outside Commodore's Cottage, but not, it seemed to me, with the same pleasant ring, and I longed to run away somewhere beyond the noise. However, I had fortunately little time to indulge in fancies. Camp gear, stores, earthenware, glass, kitchen utensils, everything must be packed before 3 P.M., and stowed on board the steam-launch in readiness to sail at 6 o'clock the following morning.

I often wonder how we got it done. I think it must have been, not only by the zealous assistance of officers and men, but by the stimulus we ourselves received from the invigorating atmosphere of sympathy and good will which surrounded us. At all events, before sunset, Commodore's Cottage was ruthlessly plundered of such of its contents as would fit our camp, and the croquet ground again stood empty as we had found it. I felt "rooted up" and miserable ; but without a doubt that we were on the right way. So, to cover my nervousness and restlessness, I went to bed.

Next morning, as the sun rose, a rare procession passed down the coast. A steam-launch, with Captain Phillimore and David on board, towed along two well-laden lighters and a sailing pinnace, and carried, more-over, quite a tail of little surf-boats, or "dingeys." The busy trade-wind had sunk almost into a dead calm, the sea seemed still asleep, everything was in favour of an easy landing, and I felt hopeful, though anxiety made the hours seem long while I waited for news. I could neither read nor write, nor did idle musing soothe me, so I made believe to mend a pair of gloves, and ever after, when I wore them, I was wont to trace the anxious thoughts sewn in with every stitch. I take some pride in glove mending, but this pair shows many weak stitches, and sad botching, just where I threw them down in disgust, and, bidding patience good-bye, put on my hat and walked into the noon-day sun.

"Sister Anne! Sister Anne! do you see anybody coming?" "No!" That movement far off among the clinker is only the rising of the heated air, trembling over the burning stones. That grating sound is not of wheels, nor is it the crack of a distant whip. It is only the morning gossip among these chattering grasshoppers. But at last, and sooner than I had any right to expect it, there was really the sound of

wheels, and good news was brought to me. Everything had been landed without a scratch, the foundation of the Heliometer House was already laid, and
the new harbour thus established, had been christened
by Captain Phillimore " Mars Bay."

On the following morning another procession wended
its way from Garrison to Mars Bay—this time by
land. It consisted of sixteen Kroomen, bearing the
Heliometer-tube, Transit and other instruments. The
Heliometer box was lashed to a mast and set out on
its perilous journey, borne on the shoulders of eight
Kroomen—four in front and four behind. The other
eight carried the lighter boxes and acted as a reserve.
Strong stalwart fellows they were, looking like so many
pillars sculptured in black marble ; and we saw them
start with something like confidence.

Soon my husband followed in the vehicle (which by
this time I had discovered it was legitimate to call a
cart), but what was his horror, on overtaking the procession, to find that these faithless bearers had unswung
the box, and were coolly carrying it on their heads.
This mode of transport looked most unsafe, and he
remonstrated, but to no purpose. " Krooboy must
carry thing on him head—he no can carry with pole—
get tired." And so the trembling astronomer was fain
to be content for the first part of the way, but when
the plain was past and the clinker appeared, his
patience gave way; he could bear it no longer. The
box was accordingly lashed to the mast again, amid
some grumbling at first, but it soon passed off, and a

few kind words made the shining black faces as genial
as ever. Then, with slow and careful steps, and with
much laughing and chattering, the precious thing was
borne over the rocks in safety, and when at last Mars
Bay was reached, its tired guardian sighed out in his
relief, " All's well that ends well."

Two busy days followed. On the third all the labour
of construction was over, and our marines were able to
enjoy their Saturday half-holiday in Garrison.

Only three days to pull down, transport, and re-erect
an Observatory, in the face of every difficulty that land
and water could offer! The seemingly impossible had
been accomplished, and yet no observing weather had
been lost; for during the three nights that the Heliometer had lain in its case the sky had been cloudy at
Garrison; while, still further to increase our satisfaction, the men at Mars Bay reported clear skies at the
New Station.

David had meantime taken up his abode at the new
Observatory, but as yet I had seen nothing of it; so,
when Captain Phillimore kindly invited me to drive
there with him this Saturday afternoon, I gladly
accepted his invitation. I was burning with curiosity,
and, as David intended to spend Sunday in the wilderness, I was anxious to see what comfort he could have.
Little enough, I knew; but with the comfort of clear
skies he would not much miss any other comfort.

It was a dazzling, dusty afternoon, and the sun was
yet shining in full strength when we left Garrison for
Mars Bay. I now saw by the light of day the road I

had followed on the night of my expedition, and I thought the darkness had covered much that might have cooled my courage. It was an ugly road, and yet well-favoured compared to what was to follow.

One object along our route I must mention, for it struck me pleasurably with a sense of freshness after a month's residence on Ascension. Actually a tree! A single palm stood erect in solitary dignity on a rocky ridge by the sea, and its grand outline was bold against the sky. A type, it seemed, of life in the midst of death, and the sight of this monarch of the East reigning alone, here on the parched desert, without a peer, without a subject even, filled me with wonder and pity. I wondered how he had sprung into life on this barren shore, and I pitied him, as Moore pities the "Last Rose of Summer." Truly this old palm was left blooming alone, and no future summer was likely to give new life to his dead companions.

No other sign of vegetation was to be seen along this road of ruts, but I espied, as an object of interest, the cairn of clinker which marked the termination of my voyage of discovery, and it showed plainly that, by keeping too close to the sea, I had run into the clinker sooner than I need have done. Captain Phillimore directed our course slightly eastward, and, amid a blinding shower of yellow dust, we drove between Saddle Crater and Round Hill, two red cinder heaps, each about 150 feet high. We then tied the horse to the first outlier of clinker and proceeded on foot. But instead of following the winding watercourse, where precipitous masses of clinker cut off every breath of wind, we took a short cut across the rocks, which, like most short cuts, proved to be longer than the orthodox way. After a mile (I thought it was three) of climbing, leaping, stumbling, scrambling, I reached the Observatory, hot and breathless, with torn shoes and skirts, and a considerably ruffled temper.

We found David, who was surprised to see us, adjusting his Heliometer, and he hoped to have it ready for use that night. This was cheering so far, but the surroundings were the reverse. I am at a loss how to convey to anyone who has not seen it, an idea of what sort of flooring clinker makes. Imagine the neighbourhood of a great iron foundry strewn with the accumulated slag of years—some of it in rough compact masses of various sizes—some reduced by the action of time into a fine powder, ready to be stirred into a cloud with every breath of wind.

Such was the ground-work upon which one had to make things comfortable at Mars Bay, and I felt very miserable to think that my husband must spend Sunday in such a place, without even the consolation of a companion to share his discomfort. However, he did not consider companionship a consolation under the circumstances, and would not allow me to stay with him, much as I wished it. So I seized upon Sam, to help me to do what little could be done to straighten matters in the short time that Captain Phillimore could remain.

There was no hope of improving the condition of the tents, floorless and curtainless as they were. The bed, chairs, boxes, &c., inclined at all angles among the rocks, in vain attempts to reach their proper level; and it seemed to me that the less unpacking done, the better for the goods and chattels, because of the fine dust which settled thick upon everything. So, after Sam had swept out the Transit Hut (which had a wooden flooring) with the stiff straw case of a claret bottle, we decided that here the bedroom must be, notwithstanding an unsightly brick pillar in the centre, newly built and smelling of mortar. It made by no means a dainty bower when our best was done, but the intending occupant was charmed with it.

After many injunctions to Sam, who was as yet the only member of the household staff here, I had to return to my drawing-room in Garrison, feeling like a Dresden China figure, and content myself with the promise that I should be allowed to come and take up my permanent abode at Mars Bay when the tents were floored and properly pegged to the ground.

I spent a too comfortable Sunday at Commodore's Cottage, and was not lonely, thanks to the kindness of my neighbours; but I had many anxious thoughts about the success of the new settlement. I feared lest in escaping one trouble we had fallen upon a worse—I doubted whether our health would bear this gipsy life under such a sun, and I felt how useless cloudless skies would be to a sick astronomer.

On Monday morning David came to breakfast with

me at Commodore's Cottage, bringing good news of
the skies, but a miserable tale of domestic experiences.
He was ill, I could see, but perhaps only from fatigue;
for he had worked almost uninterruptedly for three
days and three nights, and, instead of being able to
rest during Sunday, he had suffered from a plague of
flies, which left him not a moment's peace. His food
was so plentifully seasoned with clinker dust, that he
could hardly touch it; and the condensed water, from
standing in the sun, tasted flat and unrefreshing. An
old sprain in the knee too, felt hot and uncomfort-
able; and all this made him fear that, consistent with
health, life at Mars Bay was impossible. Rather than
risk it, he would undergo the fatigue of a journey
across the clinker every day, so as to have the comfort
of a flyless rest and a dustless meal in Garrison.

On hearing of this dismal state of things, Captain
Phillimore, with his usual kindness, gave orders that
one of the donkeys, which were loose on the clinker,
should be caught for my husband's use, and undertook
in the meantime to drive him to and from his work, as
far as the horse and cart could go. This again made
things smooth, and I confess to having felt an intense
relief in the prospect of remaining in the cottage. Not
that I feared the discomfort of the tents, but I feared
the sun, the water, and my cook. Hill had worn the
expression of a martyr ever since this move was con-
templated, and, having undertaken the journey on foot
once, vowed flatly that he would never do it again. I
had seen rebellion armed to the teeth, and desertion

threatening in the distance, so I felt helped out of a difficulty, and "thanked my stars."

But my stars were not long propitious. On Tuesday morning, when the cart from Mars Bay came in sight, there was no David, only Graydon in his place. I felt cold and sick, for I knew well what had happened before I heard the miserable words, "Mr. Gill is very unwell, ma'am, and his knee is bad." A pencil note told me that he was only very much exhausted, and would come round with the turtle boat in the evening for a few days' rest, as he did not feel able for work. But the idea of his remaining in that wretched place till evening was intolerable to me, especially as my cross-examination of Graydon brought out the fact that during the night he had fallen off his observing chair, from giddiness and faintness.

In my distress, I went to ask the doctor to drive out with me at once; but he proposed, as a better plan, to send one of his assistants alone, assuring me, that some carpenters, at work at Mars Bay that day, would carry my husband across the clinker without delay, and he could be brought back in the cart much more comfortably if I remained behind. I knew he was right, yet it was a struggle to give in, and I don't think I ever spent a more anxious morning. I was full of self-reproach. Knowing well my husband's zeal and utter disregard of his health when work was to be done, I had not urged him to rest. I began to see that over-anxiety about the "Opposition of Mars" had blinded us both, and for me there was the less

excuse, as I ought to have been attending to more
sublunary matters. I also bitterly regretted having
been persuaded to stay in Garrison during Sunday,
while David was being deprived of the rest he so much
needed, from causes that I could, in some measure at
least, have obviated, had I been with him.

It was indeed a miserable morning. The workman
sick from over-work, and the work yet to be done !
With this load on his mind I feared recovery would be
slow, and I felt mentally and physically unfit to nurse
him. Sympathy has its disadvantages, and too much
of it between nurse and patient I hold to be a misfortune. However, I did my best to be cheerful when
my husband at last arrived, and tried not to show how
much I feared that his illness at this critical time
would be fatal to the success of the work. The doctor's
visit was a great relief. On examination, the aching
knee was found to be swollen and inflamed, but not
dislocated as I had feared ; and the sickness was
pronounced to be a slight local fever called "The
Rollers," brought on, in this case, by over-fatigue and
exposure to the sun. The news that a few days of
perfect rest was all that was required gave me good
heart, and everybody's kindness helped us through
these dreary days, when, for the first time we rejoiced in cloudy nights, and there is little doubt
that they contributed much to my patient's speedy
recovery.

On the fourth day he was fit for work, and only one
clear night had been lost. But the continued lame-

ness obliged us to abandon the idea of his making the daily journey to and from the work, and we now saw that it would be necessary to live at Mars Bay, at least for a time.

Consequently, on the 10th of August, David, accompanied by Hill and Graydon, set sail once more for Mars Bay, and this time he took with him such things as his short, but unfortunate, experience had taught him were necessaries of life on the clinker. A supply of mosquito net and a water-filter he had found to be essential; and fortunately the canteen was able to furnish the former, and the hospital the latter.

It so happened that I was not able to leave Garrison that day, so I retained Sam as watch-dog, hoping he would give me the protection of his presence in the deserted Commodore's Cottage. " Oh yes, me take care of you, ma," said Sam, boldly; but alas! his courage sank with the sun, and, when darkness came on, he begged to be allowed to go to Krootown (as the Kroo quarters are called), because, " me 'fraid Fetish, ma," and so my valiant guard fled with all the speed a pair of bare heels was capable of; leaving me all night to the tender mercies of " Fetish." I remained undisturbed, however, and next morning we journeyed along the rough road that I had climbed before, to join the rest of the household; but instead of Faith and Hope, I had, this time, Fear and Trembling for travelling companions.

CHAPTER IX.

MARS BAY.

A gloomy Home-coming.—A wet Sunday.—Our tents.—"Sam again."
—Mail-day. — Setting the House in order.—"Hard-backs."—
Watchful Nights and Weary Days.—Sam *versus* Graydon.—
Scarcity of Water.—Good Samaritans.—An Eclipse of the Moon.
—Our Cooking-tent by Night.—Guests at Mars Bay.

IT was a gloomy home-coming. I was tired and
cross, and the skies were angry too. Clouds were
thicker and heavier than I had ever seen them in Garrison; and not even the news of a complete measurement of Mars on the previous night, could remove the
heavy weight of fog that had settled on me and Mars
Bay. My bright vision of a land where skies were
always blue was bidding me farewell, and the parting
was grievous. To be sure the tents were much improved since my former visit, and altogether there was
now a good foundation on which to build comfort; but
I looked at everything through the fog, having, unfortunately, lost my couleur-de-rose spectacles for the
time, and I felt that I should not find them until I had
seen Mars.

No observations were possible that night, and next
day—Sunday, it rained heavily. This was the first wet

day we had experienced in Ascension, and the first we
had ever spent under canvas. Our tent doors of course
faced windward, and a tepid shower-bath roused us
early in the morning.

The bedroom-tent was now floored with undressed
planks. The ropes were well secured to the ground
by iron pegs, driven into the clinker, the usual wooden
pegs having no hold here. An ample mosquito net
protected the bed; and a military-chest of drawers, an
iron wash-stand, a bath, and a couple of wicker chairs
completed the furnishing. Not quite, by-the-bye. I
have forgotten to mention the little foot-square mirror,
set in the remnants of a mahogany frame, and the
glass of which was certainly not an optical plane.
This exasperating piece of furniture was hung on the
tent pole, rather high for my convenience; and what
with having to keep out of the rain, and stand a-tiptoe
to catch an occasional glimpse of a distorted image,
who shall blame me if my toilet this morning was
somewhat awry ?

A walk of thirty yards over the rocks and under an
umbrella, brought me to the dining-room in search of
breakfast. The last time I was here, a large uncur-
tained marquee did duty for salle-à-manger, but now a
bell-tent had been pitched beneath it, forming a double
tent with a verandah between the inner and outer
tents, and protecting us on this wet Sunday from the
rain, as it did from the sun on many hot days to come.
Here there were no planks to walk upon; only concrete, not yet dry, and coming off sticky on my skirts.

Breakfast was laid on a deal table covered by a very damp tablecloth, and there was a sort of chilliness, even in the heat, which made the cooking-tent the pleasantest part of the establishment this morning. It also had a concrete floor, which the heat of the little stove had somewhat hardened and made pleasant to walk upon.

Sam had spent the night curled up in a packing-box, very close to the stove, poor fellow! Hill also had slept near the scene of his future labours, and Graydon's hammock swung in the Heliometer House. Rather to my surprise, and very much to my relief, I found the men contented and cheerful. All Hill's incipient rebellion had disappeared, and he was now busy preparing breakfast with his usual skill and deftness.

Let those who find long wet Sundays depressing in Scotland, pity us on this wet Sunday in Ascension, with no cosy chimney nook to take refuge in, and with all the restlessness of disappointment and expectation upon us. Lava rocks may be more comfortable to walk upon under leaden skies, and tents cooler when rain-soaked, but the sun does much to lessen the loneliness of barren nature, and I think one never feels so utterly and so miserably shut out from the warmth and geniality of the world, as when a crawling mist draws itself slowly round the horizon—narrowing, ever narrowing, till you seem to feel it creep into your flesh, and stifle you with its heavy breath. To-day it was the world of Mars that we grieved most at being shut out from ; nevertheless we rejoiced that since cloud

and mist did come, they brought rain with them, and
we hoped soon to hear of a rise in the water-tanks.

In the evening my husband read prayers to our
little household in the Heliometer House. By this
time the rain had ceased, but the mist still hung low
in the east, and the setting sun, now free from cloud,
threw a strong red light on the lava rocks immediately
around, bringing out their rugged lines in sharp contrast to the distant mist-covered mountain. The sea
dashed against the land at our feet with a surly growl,
and dark-winged birds whirled overhead, uttering
shrill cries. Altogether it was a melancholy, chilling
scene, and made one, without knowing why, think
longingly and lovingly of home, with its bright fire-
sides, and restful church-going Sundays.

Do what I would to keep my thoughts from wandering, a certain little Scotch strath *would* rise up before
my eyes, blotting out with its gentle loveliness the
wild, lonely, lava hills around us. I was no longer in
a canvas-covered tent, but surrounded by our dear
ones at home, in an old grey church, standing in its
quiet God's acre. The mental journey was very pleasant, and I fear my wandering thoughts were brought
back to reality not by any effort of their own.

A check was given to their wool-gathering by a rapid
movement of Sam's right foot in a hasty attempt to
crush some offending or unoffending insect, that had
incautiously come under his notice. The intent was
arrested, however, by a severe glance from Hill, and
Sam relapsed into solemnity.

Poor Sam! I ever look back on him as a cheerful element in our Ascension life. His oddities and droll sayings furnished us with continual amusement; and when we heard peals of laughter issuing from the kitchen-tent in the quiet evenings, we used to say to ourselves, " Sam again."

Most unexpectedly the sky cleared after sunset on Sunday evening, and a good set of observations was obtained; so the night was brighter to us than the day had been, in every sense of the word.

On Monday a crushing sense of neglected corres- pondence came upon us, and we could no longer yield to procrastination. As yet, no English mail had gone from, or come to Ascension since our arrival, nearly five weeks ago. Indeed, we had brought the last news from England with us on the 15th of June; but now the August mail day was imminent, and I had made no preparation as to letters. Not one had I written, less from want of time than from want of will; for, though not exactly like an American lady of my ac- quaintance, who only writes when she is in good spirits, I was loth to send letters full of uncertainty and worry to friends at home, who, I knew, would open them full of anxiety for good news. I could not bear to disappoint them, so I put off writing in the hope of being able to write more cheerfully, and thus I overburdened the last days before the mail.

I rather think too, that David had got into a like mess with his correspondence. At all events, pens, ink, and paper were in great request with us for three long,

hot—very hot afternoons; and much as I longed for
the mail to bring us English news, I was somewhat
lazy to give it our Ascension news in return. However, we wrote busily, and on Wednesday the 15th, our
letter-bag went off with a report to the Astronomical
Society of four complete determinations of the Parallax
of Mars. Not by any means such a favourable account
of our work as we should have liked to send, but better
than at one time we had dared to hope for; and now
that the best in our power had been done, we were
more patient to wait for the future, and almost satisfied
with the present.

On the morning after our letters had been dispatched,
David was busy with hammer and saw, making me a
work-table out of some odds and ends of undressed
planks, and I was toiling, hot and awkward, "getting
up " the first week's wash, when Hill interrupted our
labours with the welcome news, "Please, sir, the mail."
Down went hammer and saw; down went the flat-iron
and burned a hole in my pet collar.

Those unhappy people, who have the misfortune to
hear the postman's daily knock, will not be able to
realise the intense excitement and delight of mail-day
after a newsless lapse of two months. It was really
worth waiting for; every little item had gathered interest from every salt wave it had crossed, and each
home name had won a sweeter tone from each hour of
silence. How often I read these letters I know not,
nor should I like to tell how much time I devoted to
the perusal of the newspapers.

All the male population was aglow for war news, and I tried to be interested, but could feel little sympathy with Turk or Russian, while the tales told of both were so horrible, that I sickened as I read, and felt thankful that the din of battle came across the sea to us with a muffled sound.

After the excitement of mail-day was over, I set about putting my house in order in right good earnest, having been able hitherto to do so only by snatches. Outside, great improvement had already been made by our servants and a party of Kroomen. The difficulty of getting from one tent to another over loose clinker stones was at first very great, and my shoes were sadly cut and torn in the process. But now, Hill and Graydon had removed many of these stones and established a branch system of little paths running from door to door, which the Kroomen filled up with beautiful white sand from the beach. The benefits of this work were manifold. It saved shoes and feet, showed a safe path at night, and, best of all, laid the dust to some extent; for what I have called sand is not really so, but minute fragments of shells and other disintegrated marine matter, worn very fine by the action of the waves, and too heavy to be stirred by the wind; hence the advantage of burying our dust in it.

Outside our bell dining-tent, and within the shade of the larger marquee, I had the vacant space covered with this sand, and, when bordered on either side by pretty pink and white shells, it made quite a grotto-like verandah. Here too, we swung our hammock—a

thoughtful gift from the officers of the *Cygnet*—and, after
having learned the art of getting into and out of it without disaster, this nest became to me the happy scene
of—shall I confess it ? I am afraid ; for it is almost
as great a crime in the eyes of our lords and masters
(except when they indulge in it themselves), as "five
o'clock tea," and it is almost as tempting and bewitching. Have you ever tried in the tropics, O æsthetic
friend, an afternoon siesta ?

With this sand-covered verandah for protection, I
now ventured to get out books, cushions, chintz table-
covers, and other signs of civilization, and, when the
concrete floor "set," our verdict on the dining-room at
Mars Bay was, "Why ! it is the nicest, coolest little
resting nook in the world, and we mean to be very
happy in it."

A white sanded path now led to the kitchen, which
stood to the north of the dining-tent, and sufficiently
near for Hill to hear me clap my hands when I wished
to call him. To south-east stood the Transit Hut;
on a rocky knoll a little farther in the same direction,
the Heliometer House, and just below it, our bedroom ; all within a radius of thirty yards.

On pretty high ground, intersected with our smooth
white paths, and placed as we were close to the sea,
our encampment must have been an object of some
interest and curiosity to passing ships. No one else
saw us except the wild goats and donkeys looking
down from Green Mountain. It was pleasant work improving our little kingdom, although I knew we never

could make it look very pretty; still it grew in comfort if not in beauty, and we soon ceased to regret
Garrison.

The first thing, that conduced greatly to our comfort, was a wholesale slaughter of insects with carbolic
acid. Not that we, by any means, extirpated them, for
the ugly little "hardbacks" still dropped from the tent
roof upon our heads and crawled over cups and plates
unharmed and defiant of us; but they were quite
harmless, and we had no stinging pests. I did not see
more than half-a-dozen musquitoes during the whole
time we stayed here—thanks to the drought; and the
numerous many-legged creatures that we called centipedes did not bite like members of the real family—
of *these* only one was ever encountered, and that was
before my arrival. Wherever they found a quiet
shelter flies did much abound; but the Heliometer
House, standing high and full in the wind, was comparatively free from them; so there we made our study,
by day as well as by night, and if the skies had been
less uncertain, we should now have been perfectly contented with our lot.

But watchful nights made weary days, and it was
hard work to keep energy and hope alive. My husband
had the first watch each night; then I took his place
in the morning, to call him on the least appearance of
blue sky; and in this way I do not think that a single
opportunity of observation was lost. It was really no
hardship to be abroad during these lovely nights. The
stillness of the earth charmed the soul into a priceless

I

peace, while "From the door of a tent the only splendour came from the mysterious, inaccessible stars."
The cool air was delightful in its freshness, and I used
to feel less sleepy here by night than when the fierce
sun of noonday shone upon us with all his stupefying
power.

Properly speaking, Graydon was our night watch,
and he did his best to keep awake, poor fellow; but
failure on one or two occasions weakened our faith.
Sleep is a tyrant not to be conquered in every case by
anxiety about the sun's parallax.

Sam's great fear was, lest he should be told off for
night duty, and Graydon, who was rather given to
teazing him, used to threaten him with a watch sometimes. But Sam began to see the joke, and fairly
turned the tables upon his tormentor. "Well then *I*
watch for Da-da, and you go Garrison to catch beef!"
said Sam one day with a grin, knowing well a sailor's
disrelish for a long land tramp.

Indeed, not many would have liked Sam's walk three
times a-week, across the clinker, to fetch our fresh
meat and bread; no light load, for he brought the
men's rations, and all sorts of little commissions besides. But Sam had no notion of such things as time,
weight and distance. He got up with the sun, went
to bed with the sun; when the sun was hottest he ate
his "pepper soup" and stale fish, and troubled himself
about nought beside.

Over and above Sam's trips to Garrison, our commissariat was further supplied by a weekly boat.

Every Wednesday the " turtle boat" brought us water, canteen stores, &c.; but we soon found that its punctual arrival must not be depended on, owing to the rollers which so often disturbed our little bay. On the second Wednesday of our stay no boat could land, and my first inquiry was with regard to the state of the water casks. " Only water to last till Saturday," Hill told me; so I had to order all plate and glass washing, and all " swabbing out " to be done rigorously with salt water; the precious fresh fluid to be used only for drinking, and that sparingly.

I had no tea that afternoon, so I remember it as a Black Letter Day, for even Ascension tea is better than none. This supply of water had been condensed for us by the *Cygnet*, and was tolerably good for all purposes except tea-making; but experience had taught me, that no condensed water ever succeeds in bringing out the subtle " bouquet " of the cheering cup. Hence, at Mars Bay, life was robbed of one of its choicest joys! Oh, for a Scotch mist! Oh, for a babbling brook or a " brimming river " to fill my tea-pot once again! Thus I apostrophized the Fates, as I waited for the turtle boat with her water-casks. But the rollers continued, and she came not.

Next day, Captain Phillimore paid us a visit by land, to ascertain how long our stores could hold out against the besieging rollers; and he brought more than kind inquiries, for he carried some fresh meat for us across the clinker in case of our utmost need. Capt. Hammick, who accompanied him, brought me two gills of

milk in a soda-water bottle. What a recherché luncheon
we had, and how we blessed our Samaritan guests!
Another day, and again no boat, but we were well pro-
visioned—there was still water to last twenty-four
hours, and Captain Phillimore had promised to send
a mule with a small cask of water on Saturday, if the
rollers continued.

Thus my mind was sufficiently tranquil to enjoy the
beauty of the unquiet waters. Fearful and wonderful
they looked; and, the more to increase the grandeur
of the scene, that night there was a total eclipse of the
moon. The sky was not obscured, but flecked here
and there with heavy billows of dark cloud, one of
which entirely covered Green Mountain, and cast a
misty shadow on Gannet Hill. The moon was un-
clouded and threw a yellow light over the sea, turning
the white breakers livid as they lashed the black rocks
with a terrible roar.

It was truly a grand sight to look upon, but somehow it made one long for the sound of a human
voice, and as David was in the Observatory, I was
glad to turn into the kitchen-tent, with the excuse of
asking Hill to come and look at the moon. I found
him comfortably seated on an upturned cask, reading
a *very* yellow novel by the light of a ship's lanthorn,
his little bed neatly covered with a Union Jack. In
the background Sam was peacefully sleeping the sleep
of the hard-worked, curled up in his packing-case.
Altogether it was an "interior" hardly less striking
than was the landscape without—illumined then by

the dusky light of the eclipsed moon, and could I
have used the artist's brush, " Our cooking-tent by
night " would have afforded a good subject.

The following day our little bay became less excited, and towards noon we heard the welcome cry, "A
boat in sight! " Yes! there she was, tacking boldly
against the wind, and by-and-by grappling in the surf
for the buoy, which she caught with difficulty. Besides
the water-casks, she brought us two officers of the
Cygnet, who had come to say good bye, their ship
being under orders to sail next day.

Guests at Mars Bay were much appreciated, and
to-day we had quite a merry party; none the less so
that it was unexpected and knives and forks were
rather scarce. Neither did we take it amiss if our
guests turned up their plates for critical examination
before using them, and wiped out the glasses with
their table-napkins. It was simply an act of necessity
in this part of the world, and the habit grew so strong
on us that we began to be afraid that, on our return to
civilization, we might some day, in an absent mood,
offend a dainty Scotch housewife by following the
Mars Bay fashion at her table.

After a very pleasant day, we went to the beach
before sunset to speed our parting guests. They experienced much delay and difficulty in starting, owing
to the freaks of a treacherous wave, which, in the first
place, lifted the dingey high and dry on the beach, and
then, when the passengers got on board, refused to
come back to float them off! This necessitated all

jumping on shore again, and that was just the very moment the mischievous roller seized upon to bear off the dingey unfreighted. Then there was a rush, with the result of wet feet; and our adieux were mingled, if not with tears, at least with much salt water!

CHAPTER X.

A SUNDAY SCENE.

AFTER having dwelt at Mars Bay for three weeks
in perfect peace and harmony, we were beginning
to fancy ourselves a model happy family, when one
Sunday morning we had a rude awakening from our
dream. While quietly reading after breakfast in the
dining tent, Sam's black face suddenly appeared at
the door without its accustomed grin, and wearing an
expression altogether new.

"Graydon beat me—he say he kill me," were the
only coherent phrases in poor Sam's excited, broken
English; and he really looked so savage and so unlike
himself that I felt afraid of him. David was perplexed, and went at once to seek an explanation in the
kitchen.

After a short time he came back, satisfied that he
had read the Riot Act to good purpose. Sam had been
teazing Graydon, who had lost his temper and
threatened to strike him, but the little quarrel seemed

to have passed over, and we thought we should hear no
more of it. Judge then of my dismay when, an hour or
two afterwards, I saw Graydon and Sam engaged in a
hand to hand fight on the clinker! We both rushed out
instinctively, to separate the combatants, who desisted
immediately we appeared; but by this time they were
bleeding profusely, and looked shamefully disreputable.
Each, of course, blamed the other, but both were so
excited, that it was impossible to arrive at any clear
understanding of the matter. Hill's account also was
contradictory; but he was emphatically decided in
saying, that if one of these two did not go, he would,
for he couldn't possibly live in the same tent with them,
they were so violent. And, indeed, I had good proof of
their violence in the broken cooking-stove, the door
of which was split in two by a blow from Graydon's
foot, aimed at Sam.

Here was a nice Sunday morning's work! We were
extremely annoyed and distressed, and somewhat at a
loss how to act. Each servant was good individually,
and, until now, we had thought they were in amity;
but on cross-examination, it appeared that there had
been a smouldering fire for some time, and now it
had burst into a flame, very destructive to our peace.
I must say that, as soon as their passion had died out,
both men were repentant, and very much ashamed of
themselves. Graydon's contrition was quite painful, and
he seemed overwhelmingly distressed by the thought
that "Mrs. Gill must think him a blackguard;"
while Sam wandered about, with the most woe-begone

expression of face, looking like a restless spirit of darkness.

As one may suppose, we spent a very unpleasant Sunday, and all our little congregation were absent from prayers that evening—Hill excusing himself on the plea that if he came, " They two might be agoin' it again ! "

Something, of course, must be done, but as the men were really contrite and sorry for their fault, we were willing to spare them all punishment. After some thought, my husband decided simply to ask Captain Phillimore to exchange our Krooman, on the ground that he and the others did not get on well together; adding that, for our own part, we should be sorry to lose him. We agreed to say nothing whatever about the Sunday fight—that would have been a grave offence indeed, judged on the quarter-deck of the *Ascension*.

Two reasons guided us to this decision. The first was a purely selfish one. It would have been a very serious inconvenience for David to lose Graydon, who was now well trained to the work, just as " Opposition " was approaching: and in the second place, I now discovered for the first time that Master Sam had become very unmanageable with Hill, invariably refusing to do anything not quite agreeable to himself, and making quite sure of " Da-da's " and " Ma-ma's " support in his rebellion. Indeed, I fear we had spoiled Sam by making a sort of pet and plaything of him, so there was no help for it—he must go ; but not in disgrace, for we all shared the blame, and I really felt quite miserable about it.   Captain Phillimore very good-naturedly asked no questions, but took back Sam again to work at the pier-head, and gave us another Sam in his place.

Sam the second had a care-worn, reproachful cast of countenance, painfully different from his predecessor's normal grin; and I did not get to feel the same interest in him for some time, but he proved a faithful servant, and our domestic life again flowed evenly.

We often puzzled over this sudden outburst of temper on the part of our two usually quiet, well-conducted servants, but some little time afterwards I discovered the secret from Hill.   In a sudden burst of confidence, he told me that on the previous night their week's allowance of rum had arrived, and they had drank it all at once.

Such a misfortune as this could happen only under exceptional circumstances, as it is now a rule in the Navy that the daily allowance of rum must be mixed with water when served, so as to prevent the men selling it to each other, or saving it up for the pleasure of getting drunk once a week.   This wise law is by no means appreciated however, and even the poor women deplore what is for the good of their lords. One loyal wife was heard to say, "Ah! Ascension is not the fine place it used to be ; once on a time my good-man could save his grog all the week, and make hisself quite *comfortable* of a Saturday night ! "

With a view still further to guard against this
" comfortable " state of things, no intoxicating spirits
are allowed to be sold on the island; one bottle of beer
a-day may be purchased by each man—that is all, and
although it costs 1*s.*, few of the marines are able to
deny themselves the luxury.

At Mars Bay a daily serving out of rum was, of
course, impossible, so it was sent weekly in a pure
state; hence our trouble.   He is a thirsty animal, the
British tar, and in Ascension, when you ask whether
he will have a glass of rum or half-a-crown, the invariable answer is, " Well ! sir, money ain't no use
to me on this island."   Truly, gold and silver had lost
their sovereignty here, and King Rum was all-power-
ful.   For a glass of grog what was there that Jack
could not or would not do ?   And with the Kroomen,
" Bubbly-water " (rum) was equally potent.   This
tempts people to pay the men in the coin they
like best, and we must plead guilty to having done so
many a time.   Indeed, it would move a heart of stone
to see a poor fellow, who has been toiling cheerily
for you under a burning sun, come up and instead of
asking for payment, meekly insinuate that he was
very thirsty.   My husband, I know, passed a very
troublesome time with his conscience, during the
erection and re-erection of the Observatory, and it
was only when the evil came under our immediate
notice, that we fully recognized the necessity of strict-
ness in this matter of rum-giving.   Then we strongly
resolved, henceforth, to be " cruel only to be kind,"

and did what we could to prevent a repetition of the Sunday fight.

One great evil that we had to contend against in our efforts to keep the men happy and friendly with each other, was their having so little to do. However, I was pleased to find, when I tried the experiment of getting books from the Seamen's Library, that both Hill and Graydon enjoyed reading. This did much to lessen the monotony of the day for them; and David's happy thought of providing an unlimited supply of fishing tackle, and showing a keen interest in the basket, did still more for the peace and unity of our domestic circle.

On one occasion this fishing mania gave opportunity for a cruel practical joke at my expense. I had just spent a Sunday in Garrison; and when I returned on Monday, hot and hungry from the clinker, a dish of delicious filletted fish was set before me. This I concluded to be rock-cod, to which my exceeding hunger was giving an unusual relish. More than once I expressed my appreciation of this dainty dish, testifying to it still more strongly by making an excellent meal. David kept on saying how glad he was that I liked the fish, and then, as if a sudden thought had struck him, he turned to Hill and asked, " What sort of fish is it ? "

" Conger eel, sir ! " said Hill, in a tone of suppressed amusement.

"Eels!" I exclaimed in disgust; and threw down the fork which was conveying a choice morsel to my lips.

I had all the prejudice of a Scotchwoman against the
nasty things, and nothing would induce me to have
them at table, so my ingenious husband had recourse
to this cruel experiment, hoping thereby to cure me of
my fancy. But who would give up a pet prejudice
or a pet superstition without a struggle in this age
of logic and hard facts? I was not going to yield
to the first assault, so I at once declared that I
felt sick and altogether much too ill to watch for
Mars that night. "Revenge is sweet, especially to
women."

Besides this horrid conger—which, strange to say, is
considered quite a delicacy in Ascension—our little bay
swarmed with fish of every shape and size, from the
monster shark, that cost us much in the way of
lost lines and hooks, to a lovely nimble wee fish, vul-
garly called "five fingers," striped with changeful
greens, and glistening like a rainbow with every restless motion. The white-fleshed cavalhoe and the
savoury rock-cod were the staple food of our breakfast
and dinner-table; but, numerous as they were, it required no small skill to catch them, owing to the larger
number and greater greed of the hideous black "old
maids," with their double row of dog-like teeth. If
the rock-cod did not look sharp, these ravenous creatures got hooked in their stead, much to our disgust,
for they were unfit for table use.

Beautiful opal-coloured mackerel darted about in
the clear pools, and it became quite a sport with David
to spear them, for bait, with his iron-pointed alpen-

stock. Prickly sea-urchins lay curled up like balls of
miniature bayonets among the coral, and the lovely
sea anemones bloomed fair in our marine garden at
Mars Bay. Numbers of little crabs crawled everywhere above and about the rocks, where myriads of
" natives " lived and died; and from under many a
weed-grown stone a slimy cuttle-fish would stretch out
a hideous arm for a passing crab.

These salt pools, left by the receding waves among
the rocks, were beautiful natural aquariums, in whose
inhabitants we had great interest; and our daily sunset walk invariably took the shape of a scramble along
the beach in search of " fairlies."

Sometimes, instead of pools we found little islands
of white salt lying among the black rocks, testifying to
the rapid evaporation, due in great measure to the constant trade-wind. We did enjoy these scrambles, when,
after the heat of the day, our parched lungs were refreshed with a draught of fresh cool air; and, if too
great curiosity regarding some sea-creature cost us
a wetting, that but added to the pleasing excitement
of the excursion.

During this hour at the sea-side the subject of Mars
was prohibited in our conversation, and my husband
endeavoured to fortify himself for the coming anxious
nights, by banishing the cause of anxiety from his mind
for a season. But the success of his efforts was doubtful, and I have a suspicion that many more of our
wettings were caused by upward glances towards the
clouds, than by downward seekings into the pools.

As Opposition (5th Sept.) drew near, our anxiety
increased; and although we had by this time secured a pretty large number of observations, yet for
some nights previous to the important 5th little had
been done, and the decisive battle had still to be
fought.

The night of the 4th was very exhausting and unsatisfactory. Observations had been obtained in the
evening, but in the morning heavy cloud, with hopeful
though too short intervals of brightness, had kept us on
the *qui vive* until 5 A.M., all to no purpose. Then, just
as we were hoping for a little rest, Graydon sighted a
huge steamer, which we fancied might be the mail
(due on the 7th), so we had to give up all hope of
repose for this night and hastily finish our letters.
But it was a false alarm. As she came nearer we
could see that this was even a larger ship than any of
Donald Currie's floating castles, and that her decks
were aglow with red-coats. A troop-ship, of course;
and we were soon able, by the help of a glass, to make
out *Orontes* on her stern. What a monster she looked
as she sailed slowly by to Clarence Bay! I have no
doubt our little white encampment raised much speculation on board; but the *Orontes* did not deign to dip
her flag to us, as did some other vessels, more polite;
nor did she call at our little harbour, which, considering her size, was not remarkable.

We thought what a busy day they would have in
Garrison sending home convalescents, and men whose
reliefs had come, unshipping horses, mules, &c., and

we were not surprised that our turtle boat failed to
put in her usual appearance, although her non-arrival
considerably upset our domestic arrangements, and
had a fatal consequence with regard to our letters.
But this sad tale must be told hereafter in its own
proper place.

CHAPTER XI.

THE OPPOSITION OF MARS.

Suspense.—*Evening* success.—A little cloud.—Splendid Definition.— Sweet sounds.—A favourable Opposition.—The Mail lost.

MEANTIME the 5th of September has come. I could write no diary, and have not the slightest recollection of how I spent the day—unprofitably, I fear, in watching and waiting; finally bringing on a violent headache towards evening, which was less painful, however, than the excessive nervous excitement I was endeavouring to repress. To-night Mars will be nearer to us—his ruddy glare brighter than ever again for a hundred years, and what if we should not see him?

The sun had shone all day in a cloudless sky, but before sunset some ugly clouds rolled up from windward, and made me feel quite feverish. I could not rest, but kept wandering about from tent to tent like an unquiet spirit; inwardly resenting David's exceeding calm, as a tacit reproof to my perturbation. There he sat, quietly tying up photographs, softly whistling to himself, as if nothing were going to happen, and

K

then he actually smoked a very long pipe, with even
longer and slower whiffs than usual. Of course it
was affectation! But I wondered how he managed to
keep up the deception, and for the first time *fully*
believed what he had told me of having enjoyed his
breakfast on the morning of the Transit of Venus,
notwithstanding that it rained. Nominally, we dined
to-day at half-past five, and I found it hard work!

Six o'clock, and still the heavens look undecided;
half-past six, and a heavy cloud is forming in the south.
Slowly the cloud rises—very slowly; but by-and-by a
streak of light rests on the top of the dark rocks—it
widens and brightens, and at last we see Mars shining
steadily in the pure blue horizon beneath. It was
now seven o'clock, and David called quickly for lights.
Graydon, who was almost as much excited as I was,
answered with his ready " Aye, aye, sir," and in a few
minutes I was left alone in a pitiful state of anxiety
and unquiet.

How slowly the minutes passed! How very long
each little interruption appeared! The wind was
blowing lazily, and light clouds glided at intervals
across the sky, obscuring, for a few moments, *the*
Planet as they crossed his path. But at last I heard
the welcome note " All right," and then I went to bed,
leaving David to add the pleasant postscript of " Evening success " to his letters. When the letters were
finished, he gave them in charge to Hill, with orders
that they should be sent off at daybreak, and then he lay
down to rest.

I now took the watch for the morning. The first hours of my waiting promised well, but before 1 A.M. a tiny cloud, no bigger than a man's hand, arose in the south, and I called my husband to know what he thought of it. On this, the night of Opposition, the planet would be in the most favourable position for beginning morning observations about 2.30. Now it was but 12.50, and the question came to be —shall some value of position be lost, so as to give a greater chance of securing observations before the rising cloud reach the zenith, or shall we wait, in the hope that this cloud has "no followers"?

Being a Scot, and fully appreciating the motto of the Kirkpatricks, "I mak siccar," David began work at once in a break-neck position, with the telescope pointed but a few degrees west of the zenith. How my heart beat, for I saw the cloud rise and swell, and yet no silver lining below. I dared not go inside the Observatory, lest my uncontrollable fidgets might worry the observer, but sat without on a heap of clinker, and kept an eye on the enemy. Five, ten, fifteen minutes! Then David called out, "Half set finished—splendid definition—go to bed!" Just in time, I thought, and crept off to my tent, thankful for little, and not expecting more, for one arm of the black cloud was already grasping Mars.

My husband would, of course, remain in the Observatory for the rest of the night to watch for clear intervals, while I was expected to go to sleep. But how

could I ? I took up a book and tried to read by the
light of my lantern for a few minutes ; then I thought
to myself, " Just a peep to see whether the cloud pro-
mises to clear off." I looked forth, and lo ! no cloud !
I rubbed my eyes, thinking I must be dreaming, and
pulled out my watch, to make sure I had not
been asleep, so sudden was the change. No ! truly
the obnoxious cloud had mysteriously vanished, and
the whole moonless heavens were of that inky blueness
so dear to astronomers.

Mars now outrivalled Jupiter in ruddy splendour ; Orion had flung abroad his jewels like hoar-
frost ; the Pleiades glittered in such bewildering
multitude, that it seemed as if the lost Pleiad had
returned with a train of shining followers from some
other system. " Like fire-flies tangled in a silver
braid," they shone with a soft beauty ; and everywhere, above and around, myriads of stars dazzled
the night.

While my eyes drank in this beautiful scene, my ears
were filled with sweet sounds issuing from the Observatory, " A, seventy and one, point two seven one; B,
seventy-seven, one, point three six eight," &c.   Let
not any one smile that I call these sweet sounds.
Sweet they were indeed to me, for they told of success
after bitter disappointment ; of cherished hopes
realised ; of care and anxiety passing away.   They
told too of honest work honestly done—of work that
would live and tell its tale, when we and the instruments were no more ; and, as I thought of this, there

came upon me with all their force the glowing words
of Herschel—

"When once a place has been thoroughly ascertained, and carefully recorded, the brazen circle with
which that useful work was done may moulder, the
marble pillar totter on its base, and the astronomer
himself survive only in the gratitude of his posterity;
but the record remains, and transfuses all its own
exactness into every determination which takes it for
a groundwork."

Happier hours I never spent than those early morning ones under this beautiful heaven; for in helpless
restlessness I had again taken up my position on the
clinker. The night was unusually still, and outside the
Observatory there was not a sound save the gentle
flapping of the tents—like the wings of passing birds,
and the continual murmur of greeting from the waves
as they met the shore. Time passed unconsciously,
for I was giving my imagination full play, and when I
heard the Observatory dome shut, I could hardly
believe that I had been dreaming on a rock for three
hours. The awakening was as pleasant as the dream
had been. David was radiant, and no wonder! All
our previous disappointment, fatigue and anxiety were
forgotten in the good fortune of to-night, and now we
might rest.

But excitement made sleep for me impossible, and
after some fruitless attempts, I got up at 8 o'clock, and
sallied forth into the bright morning. At this time I
was fondly imagining Sam to be in Garrison with the

letter bag; so judge of my dismay, when the first
object my eyes rested upon was the young man quietly
smoking at the door of the kitchen tent! I called
Hill, and on my asking why he had so disobeyed
orders, he coolly replied, "Oh! the Mail isn't in yet,
ma'am, we'd have seen her pass." Be it confessed—I
lost my temper. After taking so many precautions to
make sure of the letters going at daybreak, to
have them all frustrated by such a gross act of dis-
obedience was too provoking; and it was in a very
cross tone indeed that I told Sam to be off at
once.

When David heard the news, he too was very indignant and scolded Hill severely; still we had good
hope that all would yet be well. But alas! the
truant turtle-boat, (which ought to have come for our
letters the day before, had she not been delayed by the
Orontes,) arrived to-day at noon, bringing us home
letters, and also the distressing news that ours had
missed the Mail. The Mail was under way as our
runner sighted Garrison, and the captain had waited
for him some time in vain. It was very annoying,
and Hill was exceedingly ashamed of himself; but
unfortunately, that did not forward our poor letters,
many of which were of considerable importance; at
least many of my husband's were. The only one
of mine I felt unhappy about was my letter to my
mother; and the thought of the five weeks of anxiety
that she would suffer on our account until next mail,
was the single drop of bitterness that mingled with

the sweet satisfaction of a favourable " Opposition," and the enjoyment of home news.

Well! it was one of those little clouds that often come on the brightest days, just to keep the brightness from dazzling our eyes, and to remind us that bad weather may come.

CHAPTER XII.

THE SEA-SHORE AND THE ROLLERS.

A FEW days after the triple excitement of Mars, the *Orontes,* and the Mail, two blood-stained travellers arrived at our encampment towards sunset, with torn clothes and limping gait. At first sight of them I felt a thrill of alarm, but was soon relieved by a familiar voice calling out cheerily, "Halloo, Gill, we have not fallen among thieves, only upon the clinker—the horse bolted with us, made too free with the road, and a big bump threw us out on the top of each other."

Here was a thrilling tale wherewith to stir up our quiet life, and after hearing it in full detail I registered an inward vow, never to drive across the clinker with *that* horse. Our friends, happily, did not seem hurt, beyond a few bruises and some slight cuts about the arms, but these were enough to stain their torn sleeves and give them an air quite touching and heroic.

Of course there was considerable abuse of our thoroughfare, and we now heard for the first time, that the day the *Orontes* was in harbour, several of her officers, with two lady passengers, had set out with the intention of paying us a visit. But the bumping had been such as to bump a wheel off one cart; and some accident, I forget what, having happened to the other, the whole party was obliged to return to Garrison without having been able to reach our inaccessible retreat. We were sorry both for our intended guests and for ourselves, and began to fear that these repeated accidents would have the effect of deterring all but the most courageous spirits from seeking us out.

It was really a terrible highway,—as liable to rise and fall as the Stock Exchange,—cutting short the breath of the unlucky voyager with each sudden descent, and further blinding him with the flying ash, which rose in clouds as the clumsy cart-wheels chased the labouring horse across the lava plains. And yet I preferred this route to Garrison to going there by sea. The outward voyage was still fresh in my memory; and had a landway, as sorely beset with dust, ruts, and rocks as this road to Garrison, led back to England, I would have suffered the dust-staining and the bruising rather than be "rocked in the cradle of the deep."

I once read somewhere of three death-bed regrets of an old sage. "There are three things in my past life that I would recall," he said. "The first, that I ever told a secret to a woman ; the second, that I ever let a day go by without bringing some good to pass ; the third, that I once took a journey by sea when I might have gone by land." And I was fully resolved that my latter days should, at least, not be burdened with this last regret.

But to return from this digression, and à propos of roadways, one cloudy afternoon towards the end of September we discovered, in the course of our evening ramble, that a rough path led across the little tongue of land which I have already described as lying to the south of us. Ever since coming to Mars Bay I had looked at this forest of lava, and wondered whether it might be possible by any means to penetrate it, and so reach the twin bay on the other side. But the needle-like rocks were not encouraging, and it required some practice in clinker walking, before we could make up our minds to attempt it. However, on the afternoon in question we resolved to explore, and set out about 4 o'clock in a spirit of enterprise, and armed with our alpenstocks.

We entered the rocky forest at what seemed the most accessible point, close to our shore, and then tried to steer eastward. After clambering a few yards, we noticed that in some places the sharp points of the rocks were broken off, and the hollows filled up with them, "just as if it were meant for a road," we re-marked, never dreaming that this was other than chance. But as we proceeded, it became clear that a road had actually been made here ; shadowy indeed, and we often lost it, but only to find it again at another turn. The discovery affected us with something of the scared surprise that Robinson Crusoe felt at sight of the foot-prints on the sand.

A sudden brightening overhead caused us to change our intention of crossing South Point this afternoon; so by-and-by we turned southward towards the sea, hoping to catch a breeze as we climbed homeward along the coast from South Pyramid. There, in the stony forest, high rocks on every side kept off the wind, and now reflected the heat of the sun to a painful degree. This made them the reverse of pleasant travelling companions; nevertheless, they were very beautiful. Wonderful too, they were beyond description, and I so longed to know something of their story that I was almost cross because they gave me no "testimony."

The monotonous cinder-heap on which our encampment was pitched roused no admiration, and but little wonder. It was plainly volcanic refuse, dull and dead; but here, where we now stood, all the grandeur of Vulcan's monuments was around us, fresh as the day he moulded them, and fantastic as the dreams of any fire-worshipper. Here and there towered aloft great red masses of lava, soft and crumbling to the touch, and of whatever form the wilful fancy of the time had shaped them. Partly covering some of the harder rocks was a soft snowy substance—of what nature I know not. Closer to the shore these colours and all sharp points disappeared, and the dark basaltic rocks stood alone in well-rounded outline.

One could see that they were swept at times to the very top by the restless waves, which were now dashing against their slippery sides with such violence as to send a shower of spray over us where we stood, some distance off, watching the grand contest between sea and land. Through some parts of the resisting wall the waves had cut passages for themselves, and came roaring under the rocky arches with a noise that made one wonder how the peaceful limpets and cray-fish could put up with it.

For the first time I saw beauty in Ascension. Grim and joyless, but grand and majestic, were these gloomy rocks, trimmed round the base with delicately-tinted coral, their sternness veiled in feathery foam. Millions of shell-fish covered the lower rocks, among which lurked lucid pools, lined with the wonderfully-con-structed homes of their habitants.

While poking at a lovely shelf of pink coralline in one of these grottos, trying to dislodge it, I felt my stick suddenly pulled from my grasp. Thinking it must have got fixed among the stones in some way, I was about to put down my hand to disengage it, when to my horror I saw some ugly slimy tentacles wind themselves round my trusty staff, which was now the prey of a cuttle-fish. There was not the slightest occasion for it, of course; nevertheless, I screamed. This was no devil-fish of Victor Hugo dimensions; but so hideous was the creature, that disgust, not terror, possessed me. David, who was at a little distance exploring on his own account, concluded that I had at last sprained my ankle—an accident he had been threatening me with for some time—and ran quickly to my assistance.

"Only an octopus! We have seen many of these before."

"Yes, but only baby ones, who looked innocent enough to be gorged with crabs ; this is a monster—a fiend!"

We stood watching him. Clearly my stick was not to his liking, for by-and-by he gradually unwound himself from it and sank sullenly down among the coral, looking, as before, like a tuft of harmless sea-weed. How I congratulated myself on not having trusted my hand under water! Had I done so, and had I been alone, I doubt not that this monster of ugliness would have cost me at least a limb, for I fear I should have lacked the strength and presence of mind to fling him off at once, before the " suckers " had seized firm hold —the only chance, I believe, of freeing one's self without hurt. David wished to secure our big octopus for future contemplation, and aimed at him a strong blow, hoping by chance to touch his vital part, but he only touched his spleen. Immediately on finding himself attacked, the creature emitted an inky fluid, which turned the clear pool dark as Styx, and under cover of this he made his escape, much to David's disappointment, and to my relief.

It was so fresh and cool and beautiful here by the

sea, that we would fain have prolonged our stroll; but the sun was getting low, and it would have been a serious matter indeed to lose the daylight, with such an uncertain path before us, and so much starlight work to be done. No one that has not lived for many weeks in a lonely corner of the earth, with no variation in its dismal landscape but cloud and sunshine, daylight and darkness, can imagine my enjoyment of this new scene; and, notwithstanding aching feet, cut shoes, and tattered skirts, I felt eager to explore further another day, and to follow the rough road to its end.

The nights succeeding "Opposition" had been wonderful, and each one filled many pages with Heliometer measures. All fear of failure having quite passed away, I felt it no treason to long for a cloudy afternoon, that we might with comfort extend our exploration into the unknown country.

It was not long before such a day arrived; and as Mars Bay was so fortunate as to have a lady visitor at this time, we made a pleasant little party on our second excursion. Only Sam was left behind; and Hill and Graydon, bearing between them a pickaxe and a basket for booty, were followed by *Beauty* and *Rover*—two important members of our household, whom I feel ashamed of never having mentioned before.

Poor Beauty was a sorry dog, with a coat lanky, stubbly, and grey; a rather imbecile expression of countenance, and a pathetic limp on a hind foot. She was affectionately attached to Hill, and so gentle-mannered, that we loved her soberly, and felt indignation against the facetious wit who had named the poor old thing Beauty. I doubt her personal attractions even in youth. But possibly she may have been a beauty in her day, just as every exceptionally ugly old woman reports herself to have been ; and, perchance, she was once lithe and nimble, like old Argus in his youth.

Such was our Beauty ; and I bring her before the curtain first, not so much with the idea of "Place aux dames," but for fear, lest, once fascinated in the enumeration of Rover's charms, I might have forgotten her.

Pretty little Rover ! To you belongs the honour of being my first love in dogs. Not that I was the happy possessor of this fascinating poodle, but I took a great interest in his education, and shared the excitement of his washing-day with Graydon, who was his lawful master. Altogether he was too delicate a "beastie" for clinker life, and it cost us much trouble to keep him from rolling his silky snow-white coat in the black dust. This was especially annoying just after he had been washed, and there was great difficulty in finding a place where he would allow himself to dry cleanly.

At last we hit upon the roof of the Transit Hut, and there dear little Rover used to be perched aloft, whining and shivering, the picture of pathos and despair. Once indeed this plan had serious consequences. All the ceremony of washing and "hanging up" had

been gone through one fine morning, and Rover was
growing *dryly* content in his airy situation, when the
astronomer bethought himself that a star must be
observed. For this purpose he proceeded to the
Transit Hut, threw open the roof with a jerk, and
down slid poor Rover on the clinker, fortunately
unhurt, but it was a great shock! and hereafter the
master of the house had to be warned of the washing-
days, and cautioned against a rash use of the drying-
ground.

This account of Rover will show how unfit he was for
our rough walk, and I strongly objected to his society
on the occasion. However Graydon undertook to
carry him in the basket when he felt tired, and we all
set out together, leaving no one to guard the Observatory but Sam and *Polly*—a most uninteresting parrot,
belonging to Hill, whose days were spent entirely in
eating and screaming.

We were able to find the path again with some
trouble, and now followed it right across South Point
to the bay on the other side, where it terminated.
Here we discovered, hidden among the black rocks, a
lovely white coral-strewn beach, larger than we could
boast of at Mars Bay; but being on the windward
shore it formed no harbour. This the chart showed
us was Gannet Bay, and we concluded that turtle
must have come ashore here at some time, and that
the path had been made to facilitate their transport to
Garrison; but on afterwards talking the matter over
with Captain Phillimore, he told us that the path had

more probably been made while the island was being surveyed in the beginning of the century.

Upon the rocks within high-water mark we noticed a quantity of curious dark-green seaweed. This we afterwards learned was available for dyeing purposes ; but not being aware of its properties at the time, we did not press it into our booty basket; contenting ourselves with a variety of the delicately-tinted shells with which the beach was strewn, and some lovely corals. Indeed, on the homeward journey the basket became so full that poor Rover could find no nest in it.

"This is just the sort of ground they find diamonds in," David kept on saying as he plied the pickaxe ; but all that our good fortune brought us was some curiosities in lava, ironstone, pumice, and one or two wonderful specimens of basalt, exactly like logs of charred wood without the bark. These, mixed with the bright coral, afterwards made a pretty rock-work at the door of our tent, besides giving me food for no end of speculation and wonder. For my Geology was only in its infancy, while my husband's was rusty from disuse ; and we too often failed to make our imperfect book knowledge explain the teaching of our "Schools and Schoolmasters."

For some days we had noticed great flights of small dark-winged birds pass inland from the sea, and as we were returning from our excursion to Gannet Bay about sun-set, the sky was dark with them. Their cries were loud and continuous, reminding one strongly of the cawing of rooks.

L

"The Wide-awakes are coming back," said the men, and we were pleased to hear it, being anxious to see their renowned *Fair* before leaving Ascension. These "Wide-awakes" or "Tropical swallows," we had been told, come here in thousands, at irregular intervals, to deposit their eggs on some rocks near the centre of the island; and the noise and bustle they make in their nurseries at these times have given rise to the name "Wide-awake Fair."

Since our arrival in Ascension these pretty birds had been absent, and we now rejoiced to see them winging their flight back once more. Many hundred miles of ocean had the brave little travellers crossed, and, wondering whether they had found the sea-voyage as wearying as I had done, I longed to throw them a word of welcome to land. Every day they continued to come in great numbers until the 5th of October, and I daresay afterwards; but on that day we left our sea-side residence for a week's holiday on Green Mountain.

It was now a month since the Opposition of Mars, and he was fast receding from us, but not before satisfactory observations had been secured on thirty-two different evenings and twenty-five mornings,—enough to give a good determination of the sun's distance when all were fully and properly reduced. It now only remained to complete the *triangulation* (or rigid connection by measure) of the relative distances of the stars of comparison. That this could be accomplished easily we had no fear, because the process, being entirely independent of Mars and of the altitude at which the observations might be made, could be continued even till the month of February if required.

During the critical part of the work we had not been sensible of the amount of bodily fatigue undergone ; but now that the nervous strain was relaxing, I could see how much my husband needed rest. As for me, I had been longing to go to the Mountain ever since I had seen a bunch of damp green ferns which had been gathered there, and so I hailed the holiday morning with a light heart.

We were obliged to take with us a good deal of household baggage, as the little mountain cottage we were to occupy was not very fully equipped, and I began to fear lest the rollers should prevent the boat from coming to our assistance.

During the five previous days they had been persistent, and for the first twelve hours their grandeur and power exceeded anything I had ever conceived. I thought I had seen rollers at their worst on the day we landed at Ascension, and again on the night of the eclipse, but these I now found were but baby rollers after all. The full-grown giants shook our little encampment like an earthquake, and the noise of their thunder deafened us. What a sight it was ! My pen is quite powerless to describe it.

They fascinate one too, these mysterious rollers, and, watching them, we enjoyed our evening stroll along the shore even more than usual. Yet, each time that a great wave rose up twenty or thirty feet high,

and came thundering along to dash itself to pieces on
the beach, I shrank back with a sort of involuntary
desire to flee the sight of the suicide.

Probably the mysterious nature of the rollers
accounts in some degree for their fascination. They
are still a puzzle to science; they still afford food for
speculative theory; and it is a relief sometimes to be
able to wonder and admire without being required to
understand. This is treason to Science, I am told,
and the ignoble escape of a weak mind from the
School of Knowledge to the Playground of Imagination!

This may be, still I loved the rollers all the more
that they kept their parentage and birthplace secret,
fancying I could see in them a dash of saucy defiance
as they sprang up, the one after the other, from the
mysterious sea.

Many theories have, of course, been set forth with
regard to them by men who endeavour to arrive at
first causes. Some have attributed them to the effects
of the moon—

> "Whom Ocean feels through all his countless waves,
>   And owns her power on every shore he laves;"

some to distant gales of wind; some to tides; others
to earthquakes; but the most ingenious theory I have
heard is that of Capt. Evans, Hydrographer of the
Navy.

In the antarctic regions near the South Pole, there
are formed in the winter huge masses of ice,—not mere

icebergs, but continents of ice, following a coast-line of some hundreds of miles. In the summer these become loosened by the heat of the sun, and doubtless detach themselves in enormous masses, the length of which is measured by many miles. Such masses, falling into the sea, displace enormous volumes of water, and thus a great submarine wave is created. This wave propagates itself northward, invisible on the surface, till, encountering a sloping obstacle, like that of the submarine side of Ascension or St. Helena, it rushes up the face of the land and causes a breaker to rise from the calm sea, having all the characteristics of the *roller*.

I like this theory, and am glad to say that I have not yet heard it explained away.

Another writer on Ascension rollers describes their appearance in language so truthful and forcible that I take the liberty of quoting his words. Mr. Webster says,—" One of the most interesting phenomena that the island affords is that of the rollers, in other words a heavy swell, producing a high surf on the leeward shore of the island, occurring without any apparent cause. All is tranquil in the distance, the sea breeze scarcely ruffles the surface of the water, when a high swelling wave is suddenly observed rolling towards the island. At first it appears to move slowly forward, till at length it breaks on the outer reefs. The swell then increases, wave urges on wave until it reaches the beach, where it bursts with tremendous fury. The rollers now set in and augment in violence until they

attain a terrific and awful grandeur, affording a magnificent sight to the spectator, and one which I
have witnessed with mingled emotions of terror and
delight—a towering sea rolls forward on the island
like a vast ridge of waters, threatening, as it were, to
envelope it, pile on pile succeeds with resistless force,
until, meeting with the rushing offset from the shore
beneath, they rise like a wall and are dashed with
impetuous fury on the long line of the coast, producing
a stunning noise. The beach is now mantled over
with foam, the mighty waters sweep over the plain,
and the very houses at the town are shaken by the
fury of the waves. But the principal beauty of the
scene consists in the continuous ridge of water crested
on its summit with foam and spray; for as the wind
blows off the shore the over-arching top of the wave
meets resistance and is carried as it were back against
the curl of the swell; and thus it plays elegantly above
it, as it rolls furiously onward, graceful as a bending
plume; while to add more to its beauty, the sunbeams
are reflected from it in all the varied tints of the rainbow. Amid the tranquillity which prevails around, it
is a matter of speculation to account for this commotion of the waters, as great as if the most awful tem-
pest or the wildest hurricane had swept the bosom of
the deep. It occurs in situations where no such swell
would be expected, in sheltered bays, and where the
wind never reaches the shore. The strong and well-
built jetty of the town has once been washed away by
the rollers, which sometimes make a complete breach

over it, although it is twenty feet above high water-
mark. On these occasions the crane at its extremity
is washed round in various directions, as the weather-
cock is turned by the wind. Such are the rollers of
Ascension, and like unto them are those of St. Helena
and Fernando Noronha."

Some Ascension observers undertake to say that
the rollers are heaviest when the sun is in the northern
hemisphere, and when storms and gales are reported
in the South Atlantic in the neighbourhood of Cape
Horn. I am not able to say whether this be so, but
from notes we made at Mars Bay, it is evident that
the rollers come up from the south, and we were able
to warn the people in Garrison some hours before they
made their appearance in the harbour.

We fancied that with rollers we generally had clear
weather, so that as a rule we hailed their appearance
with joy. But on this, the occasion of our holiday, we
were willing to dispense with them, and after having
packed the bulk of our goods over-night, to be ready for
the boat in the morning, it was a relief when I awoke to
hear only the soft murmuring of the lapping tide.
Whither had they gone? Even in anxious watching
for the turtle-boat, again and again I caught myself in-
venting fables, and forgetting to decide how many forks
and spoons should go to the Mountain, and whether or
not we ought to take preserved milk and cheese.

The question of the cheese, I may remark, by the
way, settled itself at the last moment, without any
trouble on my part. Indecision fled at the sound of

"Please, ma'am, the centipedes have made a nest in the cheese!" So the wild cats had a tasty crumb thrown to them to console them for our absence, and no cheese went to the Mountain.

Long before the boat arrived, the sea was so calm that delicate china or even a Heliometer-tube, might have been taken off in the dingey without much excitement on the part of anxious owners; and our rough consignment of household gear was soon on board, with Hill and Sam in charge, *en route* for Garrison. We ourselves remained until the cool of the evening and then walked across the clinker to meet the appointed cart, leaving Graydon, Rover, and Polly, for the time, monarchs of all they surveyed.

CHAPTER XIII.

GREEN MOUNTAIN.

WE passed the night in Garrison with Captain and Mrs. Phillimore, and set off in the cart next morning for Green Mountain, following a road which winds round Cross Hill at the same elevation above Garrison as the Captain's Cottage.

We were even now not much nearer our destination than when we left Mars Bay, but there is no choice of routes to perplex the tourist in Ascension, and we took perforce this dreary way—round the side of Cross Hill, then four miles across a barren plain, diversified by the familiar piles of clinker. Black, brown, and reddish brick-dust-coloured cinder was gathered into heaps around us, and ground into dust along our path.

Here and there the road would run for a short dis- tance parallel to the iron pipe which conveys the water from Green Mountain to Garrison, and on

passing a small square block of mason-work about
half-way, we read thereon in neatly painted letters,
"Lady Hill Tank." Tanks are placed at intervals
along the whole line of pipe to hold the reserve
water, and, if possible, to gather any that may be collected in their neighbourhood. The next one we passed
had the additional attractions of a pump and a trough,
and bore the inscription, "God be Thanked Tank."
The next was called "Travellers' Tank," and so on.
Altogether I think we passed six of these tanks before
reaching the bottom of the "Ramps." *

As we were about to commence the ascent, we
passed between two boats placed upright, with the stern
ends buried in the clinker. In this way the upper
halves were converted into rustic bowers, and a plank
of wood placed across the inside a few feet from the
ground, offered a welcome resting-place to the pedes-
trian. Here the road split into two, and on a finger-
post was written, "To Dampiers;" but as the post
was placed at an exact angle between the two roads,
and pointed quite as much to the one as to the other,
we might have been at a loss had it not been for the
very evident decision of our pony, who conscientiously
chose the steeper way.

Until now we had ascended very little since leaving
the Captain's Cottage, and after four miles of tolerably
level ground were still at the very bottom of the mountain, up the steep sides of which wound before us two

---

* A name given to the western shoulders of the mountain ; possibly
a contraction for "Ramparts."

miles of a rough and stony road. Not that the mountain was two miles high, fortunately. A screw pro-
gression, into the merits of which we had been initi-
ated at St. Helena, led us slowly upward—now north,
now south, then back again, turning and winding and
making slow advance, but affording us excellent oppor-
tunities of viewing the surrounding country, which
formed a great contrast to the brilliant panoramas
of St. Helena.

Yet it was interesting and curious. Now we could
see that what before had seemed to be simply hills of
cinder were really so many craters, some of them quite
perfect with their cups unbroken; but the greater
number were worn away more or less towards the
south-east. As we mounted higher we saw too, that
the larger hills, such as " The Three Sisters," " Cross
Hill," and " Gannet Hill," were just as fiery in their
nature as their humbler brethren. Altogether we
counted *twenty-one* extinct craters before reaching our
destination—each one once a smoking chimney no
doubt, fed from the great central funnel, on which we
could now see the mules peacefully browsing, and hear
the cocks crowing as our brave little pony toiled painfully upwards.

Unlike that of most mountainous countries, cultivation on Ascension commences at the top, but unfortunately it stops there. In ordinary times an oasis of
4000 acres decks the mountain with a green cap;
but in this season of drought, the cap had shrunk to a
mere shred, and we were very near the top indeed

before our eyes were refreshed by a glimpse of real verdure.

Stunted aloes and prickly pears appeared at intervals; that was all. But the delightfully cold wind, now rushing down upon us, banished all feeling of disappointment, and at last I was stirred into enthusiasm at sight of a little family of ferns, hiding coyly from the sun under a wild olive tree. After this, things continued to improve, and, for some yards under the barracks, the naked rock covered itself with a robe of faded green and put forth trees, under the shade of which we reached the top.

The contrast was delightful—the shade, the green, the coolness; and, down below, the hot burning desert which I could now hardly believe in. Yet there it lay, like a great map stretching out towards the sea, 2,000 feet below us; and away on the southern shore we descried our white tents gleaming in the sunshine of Mars Bay.

Near the top of the Mountain are the farm-yard and a small barracks for twelve or fourteen men who attend to the cattle, the water supply, the garden, &c. Here we found Hill, who had preceded us with the luggage, and with him a sergeant of marines, who, taking our pony by the bridle, led her forward through a stone archway, surmounted by a very rusty bell, into a really pretty garden.

I could hardly contain my delight, and yet it was no gorgeous vision that burst upon me; only seven or eight large shady trees dotted here and there along

one side of the path; and beyond these, and under
their shade, about half an acre of garden ground,
broken up into plots of young turnips, parsnips, pars-
ley, and other vegetables.

A single cocoa-nut tree, and a clump of bananas
with tattered yellow leaves, grew against the end
of the long low cottage, which stood here empty
for our use. On the side facing the east a narrow
projection was built on, and in the angle thus formed
there were—oh, joy of joys!—a few square yards of
fresh green grass. At one corner of this miniature
lawn, a patriarchal bald-headed "Pride of India"
supported a swing; while close beside the front
verandah bloomed a few roses and geraniums.

Within the verandah we found some wicker chairs
and an iron sofa; there too *grew* a small round table.
From its single support, which dreamt not of plane or
French polish, some tender green leaves were opening
to the sun; but the life rose no farther. The car-
penter had crowned the still living trunk with a lifeless head, and it was on this unique escritoire that I
prepared for the October mail.

But the greatest novelty was yet to come. The
porch-like projection which we now entered I found to
be the drawing-room, and here a bright little fire was
burning—a fire within 8 degrees of the equator, and welcome too! It gave life and cheerfulness to the rest of
the room, which contained nothing more remarkable
than six chairs, two tables, and a sofa. Behind this
was a tiny dining-room, furnished "to correspond,"

but minus the sofa ; and from both ends of it the little
house ran into bedrooms. Of these we chose the outer
one towards the south for our apartment, and with a
few additional comforts unpacked from our boxes, it
appeared to us delightfully snug.

"A gipsy's life is a joyous life," no doubt; but then
gipsies do not usually pitch their tents on clinker, and
perhaps are not so keenly sensitive to dust and centipedes as one who has adopted their habits rather late
in life. At all events, I heartily enjoyed the comfort
afforded by four walls, and began with a will to settle
down. During the process I explored the little chest
of drawers in our bedroom, and the first drawer I
opened displayed therein a disgusting cockroach rushing frantically hither and thither. Worse than centipedes ! So I shut him up, and tried another drawer.
Here I discovered a penny, and some odds and ends,
which made me feel as if I had intruded. So I shut
that drawer up too, and tried another—it was bottom-
less. Then I left the disappointing old thing alone,
and stuck to my tin box, where cockroaches and
strangers' goods did not distract me.

After having put matters straight here as far as possible, I could no longer restrain my impatience to
gather a posy—an impatience, perhaps, not unnatural
after a three months' flower-famine—so I strolled
into the garden bent on plunder, and bore off in
triumph a magnificent orange-red Hibiscus, and a
basketful of blossoms from a Pride of India. These
latter were especially grateful, with their scent of

English lilac, which they resemble also in form and colour.

The next thing to be done was to search the cupboards for glasses, and out of a stock of eight, comprising the oddest variety, I chose two of the quaintest. In these I carefully placed my treasures, and straightway felt that the four walls, six chairs, two tables, and one sofa, were converted into a drawing-room.

That sofa, by the way, deserves a few words specially to itself—a whole book, in fact, for it was vile enough for any modern hero. So high in the seat that your feet dangled helplessly in the air; so low in the back as to make you think of lumbago each time you attempted to rest; and the cushion—well, I do not *know* what the cushion was stuffed with, and my imagination lacks a simile, but it was certainly neither a pillow of down nor a bed of roses.

This afternoon the Peak was clear of cloud, and we were unwilling to lose the chance of getting to the very top, for "Garden Cottage," our present abode, stood 400 feet below the highest point. So about 4 o'clock an aged mule was led to the door for my benefit, and under guidance of a sergeant of marines, David and I set out for the Peak.

Passing through the farmyard behind, we ascended for a little way up an easy slope, which brought us suddenly on a giddy height, with sharp precipices on either side. Here on our right we could see a deep gorge run along to windward. Everywhere on our left a furrowed plain spread itself out far below and

touched the rock-lined coast; a narrow ridge, covered
with coarse grass and fringed with stunted trees and
shrubs on the leeward side, led up to the Peak, now
rising straight ahead and cutting off the east horizon.

Here was a reproduction of Diana's Peak. Another
lip of a huge crater, with its lava-sawn valleys opening
to the sea. But here no soft verdure clothed the naked
precipices—Nature hid no smiling gardens in the deep
valleys—only rocks, everywhere rocks.

The gorge dipping to windward was indeed characteristic of its name, "Break-neck Valley," for the ridge
on which we stood cast itself headlong into it, and on
either side steep rocky walls confined it. Here and
there among the crags could be seen patches of coarse
brown grass, from which a flock of sheep were attempting to feed, at the risk of their necks; and occasionally
some clumps of the much-enduring aloe stood out in
decided green, and relieved the eye.

How different had the colouring been forty years
ago, when a lady writes that, "Nasturtiums covered
the slopes of 'Break-neck Valley.'" Possibly this is
the time that the "Encyclopædia Britannica" refers
to, when it tells us that Ascension is remarkable for its
production of *green vegetables !* Now there is nothing
but rocks, bare and sterile, and in the lap of the
valley a sun-burnt windmill. Near it stands a large
octagon tank, and from our guide we learn that this
gorge yields our largest water supply.

As we paced slowly along—my mule preferring of
course the extreme edge of the ridge—he was not par-

ticular as to right or left, but it must be the edge—
we passed through gorse, blackberry bushes, wild
ginger, guavas, and other shrubs which I did not re-
cognise at the time, though I afterwards learned that
they were "poor relations" of fair stately trees whose
acquaintance I had made at St. Helena. In the shade
of these were growing quantities of ferns, and the
most beautiful stag-moss I had ever seen. It was
in some places three feet high, like a miniature pine,
and looked no mean descendant of the great genus of
Lepidodendron.

As we ascended, the cold increased, and the sudden
change of scene and temperature since morning was
bewildering. Yesterday at the Equator, to-day on
some furze-clad hill of "oor ain countree." I had
put on a moderately warm Shetland shawl at starting,
and had laughed at David, who *would* load himself
with a heavy Scotch plaid which I had brought with
me for a blanket. By this time, however, I was glad
enough to be enveloped in it, shivering the while, for
the trade-wind was blowing a bitter blast, and the sun
was hastening to the sea behind us, shrouded in leaden
clouds.

Within a few feet of the top, and surrounded by a
thicket of little trees, we came upon the dew-pond—a
cement-lined cup about twenty feet in diameter, meant
to receive and retain any moisture that might be deposited here. It was at this time too new an experiment
to have entirely succeeded; but even now a muddy
pool lay at the bottom, giving hope for the future.

M

The design was good, but bad material had frustrated
it. The cement was porous, and much moisture had
already escaped, while what remained was converted
into mud. Here we tied the mule to a Eucalyptus
tree, and climbed to the Peak on foot.

A splendid view rewarded us ! Splendid at least
in its comprehensiveness, for the whole of the little
island now lay before us, and we were almost startled
to find it so small. It seemed as if, with a good leap,
one might jump over all these apparently tiny ash-
heaps right into the sea.

Turning our eyes southward, we saw Red Hill lying
far below us ; and beyond it, to the south-west, was
another red cone, which I did not at first recognize as
Gannet Hill, our neighbour at Mars Bay. From our
low level I had looked on it as a veritable mountain ;
now it appeared very insignificant indeed, and I
looked over its head with much contempt (just, alas !
as old neighbours that have risen in the world too
often do) right away to the Observatory tents, dotted
like a flight of sea-gulls on the black rocks by the shore.

Towards the west the sea appeared very close to
us ; the intervening plain, dotted with its little vol-
canoes, being lost in the shelter of the big shoulder
of hill on which stood the mountain settlement.
To northward the dark plain again appeared, stretching here and there into pretty little white-lipped bays.
I call this *plain* in contradistinction to *mountain*, and
from the height where we now stood it certainly appeared tolerably level, with occasional ridges and fur-

rows : but I afterwards found these ridges stiff climbing, and in the innocent-looking furrows were hidden
many nasty precipices, perplexing to a mule-mounted
explorer not accustomed to steeple-chasing.

On a low spur of the mountain to the north-east stood
a solitary cottage, which, with its tiny bit of garden
ground, relieved considerably the wildness of the scene ;
especially as the ridge, stretching out behind it, was
grass-covered, and had it not been for the great red
gullies sawn on either side, one might have given
to it the sweet English name of meadow. East of this
little oasis, the fire-born wilderness gave itself up to the
most fantastic and utter barrenness, and the red colour
now disappeared from the land, giving place to a greyish
white.

The formation of this corner—the apex as it were
of the triangle into which the island shapes itself—
gave a curious impression of its having been somehow
turned upside down. The topographical characteristics
are entirely reversed, and instead of little hills rising
from a plain, as on all other sides, here deep dry lakes
are sunk in a raised plateau. At this point alone the
coast is lost to view, on account of the high ground
rising so abruptly from the shore. Thus we could not
see " Boatswain Bird Island," which, from the chart,
I knew must lie close to the leeward shore. Altogether there was a certain mystery about these bare,
bleached eastern hills and valleys, which excited a
strong desire to explore.

The Sergeant, however, was not very sanguine

about ways and means. "You can't ride there, ma'am, and you can't walk," he said; and as I certainly couldn't fly, I felt depressed, if not despairing.

Our glass could sweep the north, south and west coasts almost without interruption, and between the sea and the shoulders sloping from the great Parent Head on which we now stood, all was clinker-covered plains and clinker-built hills. Some 300 feet below us, a good footpath, called "Elliot's Path" (from the name of the admiral under whose supervision it was made) ran round the Peak; and, branching off from it, another path wound its way along a ridge running southward to a little summer house which stood on the further extremity—seeming to invite us to take pity on its loneliness, and have a cup of tea under shelter of its roof.

All this, so difficult to paint with a clumsy pen, was almost an instantaneous photograph to the eye; and well it was so, for one could not long contemplate these burning plains from that elevation without shivering and teeth-chattering, so strong and biting was the wind. Indeed I was glad to take shelter from it in a "Boat-Bower," which is placed here, after the manner of those we saw at the foot of the "Ramps." This one bore the inscription, "Ascension Day, 1876," and our guide told us that the Kroomen had had a holiday on that day in order to bring it up from Garrison. Verily these good Africans have a novel idea of a holiday! and yet do not more enlightened people often work hard—very hard at pleasure-seeking?

After a short rest here we turned homewards, leading the mule behind us, and enjoying our loiter
among the shrubs and trees. Writing in 1836, Darwin says, " On the island there is no tree ; " and in
1839, Hooker finds " Ferns the principal flora." But
I venture to mention *trees* in 1877, on the strength of
a clump of fair-sized Port Jackson willows, which
blossomed yellow in the sun, sheltered on the leeward
of the ridge ; and, of course, the Scotch fir is a *tree*,
however small it may be, though I must confess, that
the little colony in this place seemed to have forgotten
the dignity of their race.

I plucked some of the beautiful stag-moss and ferns
as we went along, and also a few of the delicate pink
blossoms of the ginger-plant, which I found hiding
away among its flat, flapping, ungainly leaves ; and as
I enjoyed the sight and scent of the flowers, they
seemed to give out to me some of their fresh life.

At the head of Breakneck Valley, the wind came
rushing up the narrow gorge with such strength and
bitterness that I could not stand against it, so I seated
myself on the brink, wrapped in shawls, to let the
breeze " play freely round me " and brace up my poor
nerves—tired and somewhat wearied out by Mars and
Mars Bay. It was very delightful. " Intoxicating,"
David suggested ; on which our guide promptly re-
marked, "They call this place ' Sherry and Bitters,'
sir." We thought the name excellent, and David
declared that it gave him a better appetite than ever
did that insidious tonic.

By this time I was rather tired of walking and proposed to ride home; but the Sergeant suggested, that instead of retracing our steps through the farm-yard, we should descend into Breakneck Valley, and go homeward by a tunnel, which leads a waterpipe through the higher ground that lies between the valley and the barracks.

Now I do not like tunnels, nor any *owly* places from which light and pure air are excluded; so, not wishing to expose my nervousness, I made the mule my ground of objection.

"Oh never mind him—Jimmy Chivas knows the island better than any of us—he'll find his way, sure enough," said the Sergeant, as he fastened the reins to the pommel of the saddle and administered a phantom kick to Jimmy, who, at the sign, fled precipitately down hill to his stable. I felt deserted, and as my husband was curious in mind about the water supply, of which as yet we knew nothing, I meekly scrambled down after him to the mouth of the tunnel. Here I should have preferred to stay and gather a basketful of the lovely little ferns and lichens which grew around its damp lips, but my guides passed quickly from the light into gloom, and I did not wish to be left behind.

At first it was not quite dark, and we could see the iron pipe crawling along one side, and the pretty green roof of moss and lichen; but by-and-by we lost this encouragement, and were plunged in utter darkness. The Sergeant marched first—I last; and I

begged for the point of David's alpenstock to guide
me, for I was continually stumbling against the damp,
clammy walls of the vault. Neither did David steer
his course quite steadily, for his "topee" was re-
peatedly knocked off, which would not have happened had he kept the centre, where the roof was
seven feet high.

Most heartily did I wish the journey over, for I
grudge even a temporary loss of the most glorious of
the Gateways of Knowledge; and in darkness I have
always a painfully acute realisation of the terrible
loneliness of the blind. And still more than light do
I love air, and this little tunnel of 200 yards chilled
and stifled me.

"How do you like it?" asked David, when we
were about the middle of the tunnel. But before I
could reply the advance-guard answered, "Oh, it's
quite easy, sir, *except when you chance to meet anything!* Coming through here, our last captain once
met a bull, which was rather awkward, as I take it."

I should think so indeed, and I quite approved of
the discretion of the gallant captain, who turned
and fled before his foe, though David suggested that
he might have made a spring and vaulted over the
beast.

"Yes," I added, "and *that* might have happened
to his head which has so often happened to your
helmet."

At last a gleam of light showed ahead, and in a
few minutes we found ourselves, much to our surprise,

in the garden of our pretty cottage home. As we emerged into light and air again, and looked back upon our mysterious path, I read over the entrance, "Sic itur ad astra. Anno Domini, 23 July, 1832." "Such is the road to the stars!" I do not believe it, nor does any one of the flighty beings who envy the "cow that jumped over the moon," and are ever longing for a Pegasus to ride along the Milky Way.

How fresh and pretty the garden looked as the sun went down! We lingered reluctantly under the shade of the darkening trees till I felt so light-hearted. that only a strong sense of matronly dignity kept me from getting into the swing, and sent me within doors to pour out tea instead.

CHAPTER XIV.

SUNDAY AT THE MOUNTAIN.

View from Garden Cottage.—"Und unter den Füssen ein nebliges Meer."—The Dairy.—Church at the Mountain.—Rest.—The Weather Gardens.—Stars *versus* Cabbages.

NEXT morning, Sunday, we were awakened by the novel sound of rain pattering against the window panes, and on looking out I saw—nothing. A dense fog surrounded us, and hid the pretty garden which I doubted not was benefiting richly from its temporary concealment.

By the time I had dressed, however, the air was somewhat clearer, and when I opened the door of our little drawing-room, a curious, perplexing view lay before me. Three paces from the door the ground went headlong down a precipice of 150 feet, covering its fall with Bahama grass, castor-oil trees and shrubs of different kinds. Then a rocky shoulder shot out from the main head, and on this stood the Mountain Hospital in a thicket of Port Jackson willows, now looking cool and fresh from their morning bath. Beyond this another steep descent went sheer down to the broad, crater-dotted plain which stretched away to Garrison and the sea, and where the sun was now

shining with a fierce light, while the mist still floated
fitfully above and around us.

Here the air felt fresh, sweet, and English-like;
there it *looked* stifling and altogether tropical. Here
the moisture was dripping from every rain-soaked
shrub, and distilling scent from the single rose-bush
on which pretty pink buds were bursting into life, heed-
less of being eclipsed by the regal red blossoms of a
neighbouring hibiscus : there, among the cinders, the
parching dust still whirled in the wind, undamped by
shower or mist, and the long-suffering Madagascar
rose (*Vinca rosea*) gave out no sweetness to the
breeze.

The whole scene gave one a curious sensation of
being in two quarters of the globe at the same time ;
and, just as I was going to call my husband to enjoy
the novelty with me, the thick mist again swept round
the mountain.   In a moment all the world below my
feet was cut off, and I and Garden Cottage seemed to
be left floating helplessly in a sea of mist.

Like Schiller's Alpine shepherd, I might have
sung,—

> "Und unter den Füssen ein nebliges Meer,
>   Erkennt er die Städte der Menschen nicht mehr,
>   Durch den Riss nur der Wolken
>   Erblickt er die Welt,
>   Tief unter den Wassern
>   Das grünende Feld."

But in my case there was no " grünende Feld."

Being shut out from the Tropics, I put on my hat
and proceeded to enjoy the English garden.   Not quite

English, however, as I was reminded by the stately screw-pine which shaded the north gable of the cottage, and by the tall aloes towering above it on the rocks behind. But the beautiful scarlet nasturtiums, which clambered up the rocks, from which the terrace for cottage and garden had been hewed, were very homely, and so were the stocks and sweet thyme growing by the door of an almost empty conservatory.

Looking down on the garden and clambering nasturtiums, stood another cottage on an upper terrace, which in its turn was over-topped by towering aloes and craggy heights. It was underneath this terrace that the dark tunnel had led us, the previous evening, across the shoulder of hill from Breakneck Valley; and with this delightful air playing everywhere around, the entrance looked, if possible, less inviting than before ; so I passed it by, preferring instead, to make more intimate acquaintance with its bright little neighbour the dairy.

Hardly less refreshing than the flowers was the sight of the milk-pans. Five starved cows gave, as it were under protest, a few pints of milk daily just now, for the use of the sick in hospital ; but it was long since I had tasted my favourite beverage, and green envy possessed me as I peeped into this pretty dairy, embowered in a splendid moon-plant whose snow-white blossoms were shedding fragrance on its roof. The lettuces and parsnips seemed to grow as I watched them, and I thought how nice a salad would look on the breakfast table. In short, I felt bewildered with a sense of profusion around, and wondered if it were really true that we had left Mars Bay not two days since.

It was indeed a change, and I was enjoying it thoroughly, when the call of breakfast summoned me indoors.

In the verandah I found Sam looking very unhappy in a thick coat with a broom in his hand; and instead of the usual responsive " Good morning," he greeted me with " Terrible cold, ma," and a shiver. Indeed it did feel cold inside the damp comfortless drawing-room, and Sam grinned approbation when I ordered a fire.

We did not expect more of church service here than we had been accustomed to have every Sunday in the Heliometer tent at Mars Bay. Consequently it was a pleasant surprise to hear that next Sunday the chaplain was to come up for morning service, and that to-day prayers would be read by the farm bailiff to the marines, who were already mustering to the sound of the rusty bell over the garden gate. When the Sergeant came to ask if we would join them, we gladly followed him to a bare, dreary-looking room, where the men were seated on forms placed by the side of a long table. Two chairs were provided for us, and the four bare walls enclosed nought besides.

After so many weeks of lonely Sundays, the pleasure of joining with " two or three gathered together in His name " refreshed us. It took us away from ourselves, and drew our thoughts entirely from the every-day work, that was almost too much to us and often refused to leave our minds free for higher things. This was the most restful Sunday we had had since coming to Ascension, and our hearts rose up in gratitude for help in past troubles, and for the bright pathway of hope laid open before us.

After service we sat lazily all the morning in the verandah, doing nothing, but enjoying to our heart's content the fresh, cool breeze which still carried just enough of damp to hold the dust in subjection, though the leaves were again rustling in sunlight. With half-shut eyes one might have dreamt of an April day in England but for one thing—the silence of the woods. No clear note from blackbird or mavis interrupted the chirp, chirp of the grasshopper, or the rustling of the wind among the banana leaves; and the canaries we only heard of, but heard not.

In the afternoon we had a quiet stroll through the " Weather Gardens"—pleasant walking indeed, after our clinker experiences, though the place hardly justified the name of " garden."

Lying on the leeward crest of the hill, which forms the western slope of Breakneck Valley, are a few square patches of cultivated ground, surrounded by a brushwood of aloes, guavas, Cape gooseberry (*Physalis*), and mulberry trees, and separated from each other by grassy paths. It is these patches that are here yclept " gardens," but in Scotland we should call them a " croft," and a very unpromising one too, notwithstanding plentiful manuring and careful cultivation. The soil was black with richness, but still too dry to yield much, except the persevering sweet potato (*Convolvulus batatas*), which was creeping diligently over one or two little fields. Some English potatoes were just showing above ground in one patch, and another was in course of preparation for planting out cabbages—all hopeful signs, resulting from the few showers that we had so begrudged at Mars Bay.

For six months previous to our arrival the soil had not even been dug, so hopelessly parched had it become; and a full sense of our selfishness came upon us, when we saw how sorely our much-abused rain was needed for the general good of the community. Still we would not willingly have given up the cloudless nights at Opposition for the sake of cabbages; no, nor for pine-apples, nor for all the ambrosia in the store-houses of the immortal gods; and, though contrite, we could hardly claim to be repentant. We promised, however, to behave more justly in future, and to think of the vegetables as well as of the stars, when we should return to the little white tents that were seen glistering far away in the distance.

From this hill-side we had for the first time a good view of the "Riding-school Crater." It stands on the western plain, and looked into from above it presents the appearance of a cone, from which a horizontal section has been cut off, and the centre afterwards slightly hollowed out, leaving an elevated circular road round the top.

These curious sights everywhere, stirred up strong desires to explore, and as we walked slowly homeward through the fog, which had come down upon us suddenly, we tried to stretch the hours of our holiday week as much as possible, and to fill them with well-planned excursions. But circumstances swept away many of our Château-en-Espagne journeys, including the visit to Riding-school Crater, which, though afterwards accomplished, did not form one of our mountain excursions.

A week, or even ten days as we now resolved to make it, is so short a time, and so many interesting things were to be seen on all sides of us, that we were obliged to give up, for the most part, very long walks among the lava, in order to enjoy with fresh mind and body the more immediate objects of interest, and that we might return to Mars Bay rested instead of tired by our holiday.

CHAPTER XV.

WHY WE HAD ONLY A GALLON OF WATER.

The Brandreth Wells.—Failure of the Spring.—Mr. Cross's happy
discovery. — A fickle spring. — How the water is collected. —
Dampiers. — Mulberry Drip. — Duck-Pond Drip.—White Wall
Drip.—Middleton Drip.

EVER since our arrival on the island, we had been
much interested about the water supply, and now that
we were at the source, we hoped to be able to learn the
parentage and history of our one gallon per day.

We had already seen, peeping aboveground here
and there, the pipe which we knew conveyed the water
to Garrison, there to be stored for the use of man and
beast; but we had seen no spring, and I was delighted
at a proposal to visit the "Wells" under guidance of
Captain Phillimore, who made himself so thoroughly
acquainted with the all-important system of our water
supply.

Starting from Garden Cottage, we again passed
through the tunnel I have already mentioned; this
time with lanthorns, which showed it to be worked out
of compact beds of cinders and ashes, and occasionally
of clay and trachyte, to which clung green moss and
lichens. Along one side, just aboveground, an iron

pipe ran the length of the tunnel, and we did not lose
sight of it until we found sun-light once more in Break-
neck Valley. Here we found the two circular wells that
contributed so largely to our daily gallon of water.
These are known as the "Brandreth Wells," named
after Lieutenant Brandreth, R.E., who came out
here in 1830 to assist Captain Bates in his anxious
search for water. With regard to the sinking of these
wells, Lieutenant Brandreth writes :—

"During twelve or fourteen months the island had
been afflicted with a severe drought, and I found bare
forty tons (of water) in store. The search for it in the
low lands had failed; the springs, or water-drips, instead
of gushing out plentifully, were scantily trickling, and
the skies were glorious, but unproductive in their un-
clouded splendour. Under these circumstances I
pressed for further experiments in boring, and fixed on
a spot high up in the mountain district, on the windward
side of the island and at the bottom of a deep ravine,
the sides of which were eighty feet in height, and
where the section showed the arrangement of the strata
to consist of volcanic matter lying on beds of retentive
clay. The clouds and mist and constant evaporation
from the sea were evidently arrested by the high land
and their moisture deposited here; and the experiment fully succeeded. At a depth of twenty-five feet
from the surface we found a spring that for the last
five years has yielded from four to seven tons daily,
and has probably averaged about five tons a day
throughout the year. The question of a supply

N

of water was thus set at rest," concludes Lieutenant Brandreth.   But the rest seems to have been
temporary.

I do not know for how much longer than five years
the Brandreth spring continued to flow, but it was not
long, and only a tradition of it remained in 1877, when
continued drought again urged the Ascensionites to
renewed efforts to find water.

Then there was no sign on the slopes of Break-neck
Valley to mark the position of the spring of 1830—the
wells having long since been choked up,—but after
some trouble Mr. Cross, the Lieutenant of Marines,
hit upon it, and found, to the joy of all the inhabitants,
that water was again flowing to the extent of three
tons a day.   Now the supply had again dwindled down
to one ton, and the fickle spring was looked upon with
doubt and suspicion.   Indeed, when we saw it I should
never have thought of calling it a spring at all.   I
should have described it rather as moisture oozing
from the mountain side.

The wells penetrate for the most part a light, ashy
soil, but fortunately at the depth of twenty-five feet
they strike a layer of clay, barely two inches thick.
Here is checked the downward course of the surface
water that has percolated through the loose soil,
till, arrested by the retentive clay, the precious drops
ooze out, and, as they drip slowly from their clayey
shelf, are caught in a cement-lined basin prepared for
them below.   The water collected in the upper well
flows into the lower one through a narrow gutter,

running along a tunnel cut between the two; and it was by means of this tunnel that we were able to get *into* the wells, and not merely to look down upon them from the surface of the ground above.

The lower well is simply a reproduction of the upper one. The drip comes from the same little stratum of clay, and is in its turn carried off by an underground pipe to a large octagon tank a few yards lower down the valley. By the side of this tank stands the windmill we had espied from the Peak. By means of it, the water is pumped up to a level sufficient to send it into the pipe which runs through the Garden Cottage tunnel to the barracks. Thus far on its journey, the Break-neck Valley water then flows into the main pipe, which has reserve tanks placed along its course to Garrison.

From the northern side of Green Mountain, and near its base, other pipes run out to join the main line; and on another day Jimmy Chivas carried me down the Ramps to explore this source of supply. Here we found two large stone-built tanks called " Dampier's " and " Bates," capable of holding respectively 500 and 312 tons of water; but these are fed by no steady drip such as exists at the " Brandreth Wells," and are dependent for their supplies on such surface water as falls on the slope above. Both tanks are uncovered, and much moisture is thus lost by evaporation.

A natural gully forms a good catch-water for " Dampier's," while a cement-paved channel, laid along the mountain side, carries the precarious rain-fall into the "Bates" tank; both, however, were dry when we saw them, and the tanks empty.

At different levels on the same slope as these tanks are three little water-drips, falling from a friendly seam of compact oxide of iron which seems to extend over a considerable area. These are the springs supposed to have been found by the famous pirate and navigator Dampier, when his vessel *Roebuck* was stranded on the island in 1701; and tradition says that he was led to this happy discovery by following the footsteps of a wild goat when he was almost dying of thirst.

The spring next below the tanks is called the "Mulberry Drip," from the solitary mulberry tree growing close beside it—the best of its kind I had seen on the island; and, on the hot afternoon that we visited Dampier's, I heartily relished some of its juicy fruit. The "Duck Pond Drip," and "White Wall Drip," are a few yards lower down, and it was quite refreshing to watch the clear drops of water trickling from the damp hill-side into the troughs, from which the overflow is carefully sent off in branch lines to join the main pipe on its way to Garrison. There St. George's tank receives all contributions, and is open for more.

One more source of supply—"Middleton Drip"—exists near the centre of the island; but analysts have pronounced the water here to be unwholesome. It tastes slightly saltish, and leaves a bitter flavour on the tongue. Partly on this account, and partly because the drip is situated in an almost inaccessible spot, no pipe has been laid from it; but the water in the open trough placed there is very grateful to thirsty sheep and goats, as well as to the mules and donkeys at large on the clinker.

Excepting the insufficient condenser, more often useless than not, the catch-waters formed by the roofs of the mountain cottages, are, I think, the only other means of supplying water to the human and brute population on Ascension; and, after our visit to the Mountain, we no longer wondered why we had *only* a gallon of water per day—we wondered rather why we had so much.

CHAPTER XVI.

TRIPS FROM GARDEN COTTAGE.

In search of Silver Ore Run.—The Wrong Road.—Empty water-courses.
—Jimmy Chivas.— The Mountain Hospital.—The Mountain
Cemetery.—The "Ingle Neuk."—Excursion to Weather Post.—
Cricket Valley.—A gorse bush.—Caught in the fog.—Boatswain
Bird Island.— Ascension game.— A novel drinking cup.— The
end of the holiday.—The wizard-wanded mist.—Mars Bay and
work again.

AFTER our curiosity respecting wells, pipes, drips,
and tanks, had been pretty well satisfied, the crater-
exploring mania seized us again; and one fine afternoon
my husband and I, accompanied by Mrs. Phillimore,
set out for "Cricket Valley," one of the mysterious
basins lying among the plateaux on the eastern part of
the island. David had walked there the day before,
and had picked up in one of the furrows or *runs* near
the valley, some beautiful specimens of carbonate of
iron, the bright sparkle of which has given to the
spot where they are found, the name of the "Silver
Ore Run."

It was this silver that we now went in search of.
Making use of Jimmy Chivas alternately, Mrs. Phillimore and I followed David round the north side of

the mountain along what we confidently supposed to be Elliot's Path. But we now found, that when Elliot's Path gets round to the north side of the mountain, it becomes Rupert's Path—in honour of some other admiral, no doubt—and with the effect of needlessly complicating the geography of the little island. However, it is the same narrow mountain way with a new name; and along Rupert's Path we now proceeded towards Cricket Valley, admiring as we went the rock-hewn gulleys sliding from our feet down to the plain below, and hiding in their deep gloom, aloes and banana trees.

By-and-by we struck off northward from the main peak, across the shoulder on which stands North-east Cottage. This tiny dwelling, intended for a shepherd, was tenantless, and was by no means so bright and homelike as it had appeared to us in the distance. The broad ridge behind it could no longer, by any stretch of imagination, be called a meadow—still it was grassy almost as much as stony, and that was an advantage not to be despised. On the south side of this ridge two furrows ran eastward. Here I saw our guide look doubtful, and he ended by choosing the wrong path. Before we got within sight of Cricket Valley, a drop of several feet told of a former waterfall, and convinced David of his mistake. It was the *other* run that contained the " silver," which it was now too late to seek for.

The object of our walk had thus escaped us; nevertheless we enjoyed seeing what we had not come to see

—possibly all the more that we could season our enjoyment with a good grumble.

All along the now dry run we were much struck
with the evident signs of water in every direction.
Not merely passing torrents of heavy tropical rains,
but permanent mountain streams, strong and rapid,
must have flowed here in former days. At the foot of
the water-worn rock, which pointed out our mistaken
way, lay a large grey stone, hollowed out in the centre,
where the falling water had swirled against it; and
along the sides of the smooth-bottomed course, where
the bare edges of strata were exposed to view, we
saw layer upon layer of disintegrated lava and ash,
which had been swept down from the higher ground
and converted by the powerful agency of running water
into a kind of soft sandstone, or a loosely compacted
conglomerate.

How fire and water do vex and disquiet our wavering
earth rind; but, like the disturbances of armed Europe,
they serve to maintain the balance of power, and keep
us from sinking to the dull level of peace and
plain !

This baffled excursion was a delightful one. As
we again wended our way round the mountain side,
the setting sun, shining through a slight haze, showed
the hills and valleys between us and the sea in a soft
pink light, which varied in intensity as the fog rose or
thickened, and gave an air of gentle mystery to the
red symmetrical cones, that had looked at mid-day as
if they were fresh from a turning-lathe. Almost in

darkness we reached our verandah, glad to follow the sun to rest, and to prepare for another day's ramble.

It was now Thursday, the sixth day of our holiday, and we devoted its cool hours to visiting the little hospital which stands on the terrace next below the one occupied by Garden Cottage. Straight down the face of the hill, a short-cut footpath runs between the cottage and the hospital, at an angle of 40°. I did not fancy it, and we chose rather the more easy cart-road by the Ramps, taking Jimmy Chivas to help us to get up again, or if need be, to show us the way; for Jimmy had been forty years on Ascension, and thoroughly understood its geography. Taught by the hard experience of hunger and thirst on the clinker, when a superabundance of horses or a scarcity of food had caused him to be turned out to cater for himself, Jimmy Chivas could find the mountain stable from every crater, gully and precipice on the island; and moreover, to him was accorded the dignity of " oldest inhabitant."

But a still more distinguished mule than Jimmy toiled daily up the steep Ramps in the mountain team—a historical animal—the mule on which Sir Garnet Wolseley rode into Coomassie. When he was pointed out to me, I regarded the hero pityingly, as one that had seen better days and ought, if every one had his due, to enjoy perpetual clover and an unharnessed old age.

During that Ashantee war in which our mule played

such an honourable part, the mountain hospital had had its twenty iron bedsteads fully occupied; but at the time we visited it, the wards, both officers' and men's, were empty; and I almost said, "What a pity!" Not that I wished the healthy sick, only that those who *were* sick might have had the benefit of such a large airy room as this, with everything fresh and sweet-smelling around, a constant temperate climate, and the perfect quiet ensured by the isolated situation.

On a hill, apart and alone—with no dwelling near it save the quiet dwelling of the dead—the hospital is surrounded by its prettily laid out garden, now much burnt up, and principally dependent on Port Jackson willows and Madagascar rose-bushes for its verdure. Dotted here and there about the verandahs are a few green tubs from which tiny mignonette leaves were peeping up; and in front of the principal door two handsome aloes reared their stately stems, protected at the base by a goodly forest of brilliant prickly-edged leaves.

Having inspected the hospital and its surroundings with much interest, we then visited the cemetery which lay close by, hidden in a thicket of willows. Most of the graves were simply square, stone-built mounds bearing no inscription; but five of them had black wooden memorials at their heads, telling the sad tale of young manly lives cut off in their prime by yellow fever. A small white marble cross marked an infant's resting-place; and another larger cross,

formed of the island lime-stone and resting on a
roughly-hewn block of lava, bore a name well-known
and loved in Ascension.

I had not thought so sweet a spot as this willow-
shaded rock could have been found for the dead in
the dusty, windy, sun-baked island; and, had the
graves been turf-covered instead of stone-covered, and
the paths been strewn with daisies instead of cinders,
one might have been content to sleep the last sleep
thus far from home.

The other cemetery, near Garrison, we had not
visited, but I retained a painful recollection of the
ghastly white tombstones on the slope of Cross Hill,
and they certainly contributed in no small degree to
form my dismal first impression of Ascension. "Well,
what does it matter? As well there as here," say
Philosophy and Common Sense; and Sentiment meekly
bows her head to the wisdom she cannot controvert,
hiding in her heart the while a longing for a sheltered
grave, and whispering a prayer to mother Earth to
deck the dead with her choicest flowers.

Much to our regret, so far as exploring was concerned, Friday was a lost day. The fog was persistent, entirely shutting out all view, and occasional
showers drenched the long grass and brushwood along
the edges of the narrow mountain paths. So we had
to give up our intended excursion to the Riding-school
Crater, and content ourselves by the "ingle neuk,"
where a nice at-home feeling took possession of us,
giving extra relish to books and work.

David had gone to Garrison the night before, to see his two months' packet of letters safely off to England, and I was very much relieved when he returned at mid-day with the news that the Mail had gone—this time with our home letters snug in her post-bag. His other news was not so cheering. While enjoying a swim in the large salt-water bath at Garrison, the unaccustomed motion had again strained his weak knee, which was now slightly inflamed, and needed rest.

This accident unhappily deprived him of a share in our next day's excursion to Weather Post—the culminating point of the high ground to eastward,—where we hoped to be able to see Boatswain Bird Island, as well as the little bit of coast not discernible from the Peak. It was very disappointing, but this being *positively* our last day (except Sunday), and not altogether foggy, we could not afford to spend it by the fireside. So Mrs. Phillimore and I decided to undertake the expedition by ourselves, and set out, under favour of a cloudy sky, guided by the Sergeant, and again assisted by Jimmy Chivas.

The first part of our journey was what I have already described in telling of our search for Silver Ore Run, round the north side of the mountain by Rupert's Path, across to North-east Cottage, then along the ridge behind it, this time keeping Cricket Valley well on the right. Looking at it from the Peak, I had flattered myself that this high ground,

which is not very much lower than Weather Post itself,
continued the whole way; but a closer inspection
showed me that the ridge behind North-east Cottage
sweeps down into a deep valley at the foot of Weather
Post, and before we could ascend, into this we must
first descend, at an angle of 30°, with loose stones
following at our heels.

Jimmy Chivas was unburdened, and, with our guide,
led the van, while Mrs. Phillimore and I followed at
two points slightly apart, neither of us wishing to send
a rain of stones upon the other. These stones were
mostly detached pieces of trachyte, many of which
were covered with a beautiful saffron-coloured lichen,
—the only vegetation to be seen, until we came into
the lap of the valley. There we found a few plants of
the thistle-like Mexican Poppy (*Argemone mexicana*),
and one or two bunches of *Polypodium trichomanoides*,
hiding among the larger stones.

At the bottom of the ravine we were still high above
Cricket Valley, which now lay directly below us on the
south; and we were glad indeed to have our curiosity
regarding it somewhat satisfied, without the ordeal of
a further descent and semi-suffocation among its sun-
baked rocks.

Darwin says, "The longer axis of Cricket Valley
is connected with a north-east and south-west line of
fissure, and is three-fifths of a nautical mile in
length."

Its sides appeared to us nearly perpendicular, and
at least 400 feet in height, except at one point where

descent is possible—*easy*, I was told—but this I would rather not assert without the authority of experience. Not a scrap of verdure clothes the barren slopes of this deep-lipped basin. Some patches of wild tomatoes grow on the flat ground at the bottom, which is so level and so roomy that one might suppose Cricket Valley to have received its name from its natural adaptability to the exercise of our favourite English game. But it is due to another characteristic of the place — one that is readily discovered by any-one who has walked though it in the noonday, and listened to the noisy chirping of its myriads of crickets.

Scrambling up the side of Weather Post, we met with no sign of vegetation, except some specimens of the curious club-moss (*Psilotum triquetrum*), until we came very near to the top. Then my eyes were glad-dened by the sight of a magnificent gorse bush in full bloom, and I proposed to rest under it, until the Ser-geant should return for me with Jimmy, after having taken Mrs. Phillimore to the top. But I saw the fog coming, and soon followed on foot. Too late how-ever! it was on the top of Weather Post before me. This was annoying, and we could do nothing but sit down and wait, with what patience we might, for the chance of a peep.

The Weather Post, about 1,000 feet lower than the Peak, is of a peculiar form, somewhat resembling that of a saddle — indented in the middle and ele-vated at both ends. I did not enjoy my seat at all.

The fog was so thick that "you might have cut it with a knife," as we say in Scotland, and our hats and gowns were soon drenched. More than this, the feeling of being so utterly shut out from all that was bright and sunny made me "eerie." But just as we had decided to go, fearing the danger of sitting longer in this vapour bath, the white coast-line glimmered in sunshine, and the little hills and valleys suddenly shook off their thick covering. Yet I looked in vain for Boatswain Bird Island. It was still invisible.

When the mist cleared, we found that we had taken up our position in the seat of the saddle, with the raised ends to right and left of us ; so I climbed some little way through scrub and loose stones to the northern and higher elevation, and then the mysterious bit of coast was revealed. Here was Boatswain Bird Island at last, about a furlong from the mainland, with the sunlight streaming white upon it, while we were still in gloom—a bright picture set in a dingy frame. A miniature islet indeed—a rock, and nothing more, treeless and flowerless, but I cannot say lifeless, for the thousands of Boatswain Birds and other sea-fowl which hover about it, whirling and screaming, give a semblance of life to the whole mass.

As we stood enjoying the strange scene, the fog suddenly blotted it out, and, fearing to lose our way should the cloud grow denser, we hurried down, so as to get over the worst part of the road as quickly as

possible. By careful choosing we found it easier than
when outward bound, and, with the help of Jimmy and
our alpenstocks, we soon reached once more the ridge
behind North-east Cottage.

Here, for the first time, I saw Ascension guinea-fowl.
These are now very scarce, although at one time, it is
said, as many as 1,500 have been shot in a season.
Indeed, notwithstanding much care, Ascension can
hardly be said to be famous for its game. Some goat
stalking there is—nearly as arduous a sport as chamois
hunting, with much less reward; rabbits too, are to
be found occasionally; and wild cats, which I believe
do not come under the game laws! A short time
before our arrival partridge shooting commenced, but
some lucky man having shot a brace, this was more
than the preserves could stand, so the shooting was
closed, and my husband's gun lay unused in its case.
Had he been here now, it would have provided a
dainty dish to set before a couple of famishing
ramblers. How hungry I was! and thirsty! So
thirsty, that the Sergeant fetched me, from the tank at
North-east Cottage, a grateful draught of water in an
earthenware jug, which bore the rather startling inscription, "Brigg's Dipping Composition, measure
for ten sheep!" but I shut my eyes and drank.
In strange countries one had best have no pre-
judices.

The mountain path was tiring, especially with the
depressing atmosphere of fog which was now worse
than ever, and we thoroughly enjoyed a lazy after-

noon, chatting over our morning's excursion and re-
tailing all its little incidents to David. He had been
busy, in our absence, putting the Barracks' clock to
rights ; and, being much relieved of the pain in his
knee, felt rewarded for his self-denial in the matter of
crater-climbing.

Another Sunday, and our last day at the Mountain. Foggy too, and showery, so that we were not
able to go to the summer-house, where, with our
books, we had purposed to spend the afternoon and
enjoy a fresh view. On such a day as this, every
place was alike viewless, and the preference was
certainly in favour of one that could be reached with
dry feet ; so we remained constant to the verandah of
Garden Cottage.

Much was left undone and unseen, but such rest
and pleasure had this one week of change afforded us,
that I hoped another gap might be found somewhere
in the work before its completion. I hoped this, yet I
did not lean very heavily on my hope, knowing how
slight was its foundation. Could the Heliometer have
been transplanted to the Mountain, that would have
been indeed delightful ! But better to have left it
behind in our dear misty Scotland than to take it up
among the clouds of Ascension ; and, remembering the
happiness and satisfaction that Mars Bay had given to
us, my feelings of gratitude revived, and reconciled me
to return.

Green Mountain was the "bon camarade" who
laughed and jested with us in our hours of idleness

and mirth. Mars Bay was the tried friend who was
with us in work and anxiety, and whose bright skies
had helped us to the happy haven of "Something
Done." Back, then, to the good friend, and without
regret, only hoping for another frolic among the clouds
at another time.

On Monday morning we started homeward under a
pouring rain. The mountain team preceded us with
our luggage which was protected from the wet by a
tarpaulin, and on the top of it lay quite a little
garden of radishes and turnip-tops. They were still
drinking in moisture, in an honest endeavour to
keep fresh until they should relieve the vegetable-
starved Garrison with the first green food they had
tasted for many months — green turtle of course
excepted!

We followed in comic procession. David mounted
on "Lucky," a handsome grey donkey; while I, with
my big umbrella and numerous shawls waving in the
wind, made poor Jimmy Chivas look as if a balloon
were preparing to rise into the air with him! He
evidently did not like the situation, for his antics in
going down the Ramps slightly heated me. Nevertheless, the ride was an enjoyable one, especially when
the rain ceased, which it did soon after we started, and
the low clouds played all sorts of fantastic games with
the little red hills below.

During this visit to Green Mountain, we had never
ceased to wonder at and admire the marvellous
effect produced, when objects at a distance are viewed

under a strong light, while all immediately around
lies in subdued shade. To some extent we had seen
this in the Highlands of Scotland ; but there the sun
is so much less powerful, that the force of contrast is
not so strong and the mist-clouded landscape is less
weird. But everywhere when such a scene presents
itself, the same thought strikes one—how tame and uninteresting would all scenery grow, without the fitful,
wizard-wanded mist ! How tired would the eye become of the most beautiful landscape lying in perpetual
sunshine, with a horizon ever well defined ! But let
the veil-like mist come, and all is changed. What was
plain before, is now bewildering and bewitching—what
was prose, is now poetry; the hard outlines become
soft and mysterious, distance is felt and the landscape
is enriched by atmospheric effects.

So it seemed to us during our morning ride down
the steep side of Green Mountain. In coming up, we
had seen a sharply-cut sea-coast edging a barren plain,
which lay dull and tame under monotonous grey clouds.
Now the scene was changed. Sea, sky and plain were
lost in each other—changeful in form and colour. At
one time the thin mist, floating in detached masses
over the sea, flecked the blue water with fleecy clouds,
making it seem as if sea were sky; and the horizon
line losing itself on the mist-chequered land, made the
little ships in Clarence Bay appear to sail, phantom-
like, among the clouds.

By the time we had descended to level ground our
fairy pictures were fading fast under the light of a

---

glaring sun, and, before we reached Garrison, whirling dust and parching thirst chased away the romance
of the morning.   But only for the time—she followed
us with fairy footsteps to Mars Bay, and was with us
in our tents that night, banishing fatigue and giving
new zest to work.

CHAPTER XVII.

MARS BAY WITHOUT A COOK.

FORTUNATELY the clouds did not follow us, and the
triangulation of the Mars stars now advanced apace. A
week of lovely evenings, and fresh strength to use them
—what more could the heart of astronomer desire?
Alas! this was Mars Bay, and not Arcadia! I cannot think why the poet says,

> "Man wants but little here below."

It seems to me that man, and woman too, wants a
very great deal; and the beauty of the universe and
the contemplation of the glory of far-off worlds, what
consolation do they give, when the kitchen-chimney
smokes, when a tooth aches or a new shoe pinches?

"Is life worth living?" asks one of our modern
philosophers, and another answers, "That depends on
the liver." True, oh Punch. Then why should we
sneer if a man fears indigestion as a mortal foe, and

the heart of woman fails when a good cook falls sick in the desert where there is none to replace him?

Such piteous case was ours! Hill fell a victim to severe rheumatism at this time, and woe is me! I had never attended cooking classes at South Kensington.

Although we did not know it at the time, our cook had been originally invalided here for this very malady, caught, like every evil thing, on the Gold Coast; and the damp air of the Mountain had unfortunately brought on a relapse. He struggled with it bravely for some days, hoping that the dry atmosphere of Mars Bay would undo the mischief. But it was of no use; the pain grew worse and worse, and within a fortnight after our return from Green Mountain he was in hospital. We were sorry for Hill, and sorry too for ourselves—there was no other cook on the island, and I was doomed to the " galley."

I have before explained somewhere, that " galley " is the nautical equivalent for " kitchen; " but our galley bore about as much resemblance to a bright, pan-bedecked kitchen, as a rubbish heap bears to a trimly kept garden; and my appetite was not improved by frequent visits to it.

I have often wondered whether a certain amount of dust and untidiness be conducive to good cooking, and on the whole I am inclined to think that they are. At least, on the few occasions that I have had a man-cook, the messes have been more satisfactory than when a female sovereign held sway o'er pots and pans, but *then* I preferred not to see the process nor the sphere of action, as that materially affected my enjoyment of the result.

And, besides good dinners and dirty pans, there is another fact that I have observed in connection with male cooks—and it is that dinner is always ready and does not suffer from delay. The unexpected arrival of the hungry master, or his unpunctual return from work, was wont to throw my tidy little Scotch cook into a fever of excitement. The result was dire! stony-hearted potatoes, or potatoes of pulp half-dissolved with weeping over their unhappy doom! But to these untidy, careless ones, things seem more accommodating, and I am puzzled to find the reason. I should have been right glad to know it when my cook at Mars Bay fell ill, and I found myself suddenly called upon to make turtle-soup and condensed milk puddings—two branches of culinary art which had been omitted in my education. Nor had I learned how to keep my temper over a cooking-stove with a tropical sun overhead, while centipedes and cockroaches disported themselves in pots and pans, and the lively breeze kept up a playful game with the clinker dust among my cups and plates. Faugh! I felt half tempted to condemn my poor husband to " Crosse and Blackwell " for the rest of our stay. But anything is better than failure, and by Sam's help I got over the prime diffi-culties of the first days, at the expense of considerable loss of temper perhaps, but that I try to consider as so much capital invested in the good cause of sympathy with irascible cooks in the future!

Unfortunately for astronomy but fortunately for gastronomy, some days of cloudy weather soon set in, and the galley became more tolerable for the poor slaves. Sam grew quite elate at the trust reposed in him, and if his efficiency did not equal his zeal on all occasions, the merriment caused by his mistakes prevented our digestion from suffering by them.

One habit he had of which I found it very difficult to cure him. I would say after breakfast, "Now, Sam, this joint will require about two hours in the oven."

" Vera good, ma'," says Sam ; and when I go into the kitchen, perhaps an hour afterwards, I find the meat slowly cooling on the table, and Sam stretched peacefully on the floor, with his head enveloped in a red pocket-kandkerchief.

" Put him in little while—take him out—put him in 'gin now," is the answer to my remonstrances, Sam being fully persuaded that if the meat be done, the time of " doing " is immaterial, or, if there is an advantage, it is in being done by instalments !

Well, Sam and I did our best ; but three days a-week he had to go to Garrison for fresh provisions, leaving me on these occasions maid-of-all-work, and I was very thankful indeed when a private of marines under-took to come and " do roast and boil " for us. Then we got on smoothly, as I had no longer to stand over the kitchen fire, but could prepare puddings or tarts in the dining-tent coolly and comfortably. And one very good result was brought about by this *contre-temps* in the galley.

I have said that Sam's favourite position was lying stretched on the dusty floor with his head wrapped in a handkerchief, but on one occasion I found his head lacking its wrapper, and Sam busily engaged with a book. I was rather surprised, but had a suspicion that Master Sam was acting a little part and trying to pass for a " scholar," as his countrymen are so fond of doing. But, watching him quietly as I rolled out the bread, I began to see that he was really engrossed in his task, and evidently puzzled by it.

"What are you reading so busily, Sam ?" I asked.

" Child book, ma'—me want to learn."

And I actually found the poor fellow struggling with " Jack is a good boy," " Tom is a bad boy," and other brilliant specimens of rhetoric, such as are to be met with in the early stages of learning the English language. He read me a short paragraph, slowly and jerkingly, but without a mistake, and could spell a few words of two letters.

I asked who had taught him. " No one, ma'—me learn meself—me want to learn."

The answer affected me strangely, and humbled me. I felt convicted of such exceeding selfishness—such neglect of those around me—so ashamed of my want of sympathy with this poor negro, whose perseverance and striving after higher things placed him at once on a level with all men who are toiling with him honestly along the highway of knowledge in pursuit of truth.

Sam, with his "child book," demanded more of my respect than I can yield to any man, however highly placed in the world, who lets the God-given soil of the mind lie untilled, and knows not that "divine discontent" which stirs the soul to work out its own perfectness, and to strive always onward and upward to the feet of its Divine Creator. In little as in much, the struggle is ever a noble one, and to see it in a neighbour whom you have hitherto looked upon as an inferior, teaches a rare lesson of universal sympathy and toleration.

In all humility I now took Sam, who had already been my teacher, to be my pupil, and, I would venture to hope, it resulted in our mutual gain and satisfaction. It was uphill work at first, and I found much difficulty in conveying to my pupil any notion of sound.

Sam's idea was that c-a-t might just as well spell "dog" as "cat." His ear could detect no difference, but his eye was more discriminating, and his memory excellent. Moreover, it was a foreign tongue to him, and very often the words, when uttered, conveyed no idea to his mind.

One day I asked him what "word" meant? "Me not know," was the answer. Then I pointed to "sheep," and said, "There, that's a word," and after that Sam always *would* insist on calling "sheep" "word," or *vice versâ*. This was partly my fault, but as I improved, Sam improved, and we soon got on rapidly, notwithstanding that some time was necessarily lost by Sam's constant query, "That right,

ma'?" and his contortions and exclamations of delight
with himself when he found the guess a happy one.

How full the days grew at Mars Bay! Amateur
cooking after breakfast, followed by Sam's lesson for
an hour; then my own lesson, for David was giving
me his help towards learning something of geology.
After this there was needlework, ironing, or some
needful "tidying" to be done, before our invariable
2 o'clock luncheon of lime juice and biscuits.

In the drowsy afternoons I would read aloud for an
hour or two, something lively and interesting—this
was our daily bon-bon, and it daily grew sweeter.

Besides this, I had a good deal of copying to do.
All the observations had to be written out in duplicate
for transmission to the Royal Astronomical Society by
the next mail, and it was necessary for me to do part
every day so as to keep abreast with the night work.
Then the evening walk on the beach, letter and journal
writing, my own reading, and occasional visitors
filled the short days to overflowing, and very happily.
Had it not been that we needed the morning hours to
make up the sleep lost in star-gazing, I think we
should not have let slip unused the two most charming
hours in a tropical day—from 6 to 8 a.m.; but as the
summer advanced, an increase of flies and the intense
light in the tents made a noontide siesta no luxury,
and we were wont to be sad sluggards in the morning.

Since the departure of the *Cygnet* late in August, no
ship brought fresh faces to cheer us until the arrival
of the *Boxer*, which came into harbour towards the

end of October. Then for some weeks we were
enlivened, even at Mars Bay, by occasional visits from
her officers, whose pleasant society was a delightful
variation in our life.

On one occasion a party of them, with one of our
island officers, were unfortunately overtaken by rollers
on their way to spend a day with us, and landed at
Mars Bay with great difficulty, looking strangely white
and woe-begone. Was it fear? Impossible! Was
it sea-sickness? I dare not say. The doctor confessed to it, to be sure, but with the others it was
biliousness—the sun—anything—everything—*except*
sea-sickness. A few hours however on *terra firma* and
the alarming symptoms disappeared; but the rollers
continued throughout the day, and our guests were
persuaded to return to Garrison by land.

The little bay was in a whirl of unrest, and when
the coxswain, who had brought the party ashore in the
dingey, was sculling out again to the boat, he had a
narrow escape with his life, poor fellow. At least so it
seemed to me, although the sailors were less excited by
an event of such common occurrence on this surf-
beaten shore.

During the income of the rollers, there occur short
periods of comparative calm, at intervals of ten
minutes or more, which must be taken advantage of
skilfully, to escape to the smooth water beyond. But
our coxswain, less experienced in "roller-work" than
the Kroo boys, was impatient and pushed off a little too
soon. The consequence was that a great wave, rear-

ing its curling crest to sea-ward of him, and breaking before it reached the shore, poured itself into the dingey, which was upset in the twinkling of an eye, and the coxswain was washed into the bay.

Graydon, who was standing on the beach, rushed to his rescue, and then for a few moments both disappeared. I became breathless, and thought of sharks with a chilly dread. But it was only the capsized dingey that had got in their way; and presently the men scrambled ashore, none the worse for their adventure, except that Graydon had a pretty severe cut on the shoulder, where the dingey had struck him.

The *Boxer* had come to us *viâ* St. Helena, from the Niger Expedition, for provisions and repairs, and to recruit her fever-stricken crew. A short time before, she had been employed in the blockade of Dahomey, while the king was preparing his indemnity of palm-oil for offended England; consequently her sailors knew too much of surf and rollers. During the palm-oil season, I am told that the average loss from these causes, on the coast of Dahomey, is a native a day, despite their ingenious surf-boats and the rare skill of their brave crews. How many of our luxuries are truly "lives o' men!"

CHAPTER XVIII.

WIDE-AWAKE FAIR.

THE *Boxer* remained with us until the 17th of November.

Some days before she left, David and I set out to Garrison one afternoon, on pleasure bent, and found the croquet ground, the aforetime site of our Observatory, converted into a lawn-tennis ground. Generally speaking, at 4 P.M. Garrison is dead, to all outward appearance—the sail-cloth blinds are still drawn round the verandahs, and nothing of life stirs abroad. But to-day there was life, without doubt, in front of Commodore's Cottage, and it struck fresh and charming upon us, in contrast to the solitude we had left behind.

Captain Alington of the *Boxer*, who had for the present taken up his quarters at Commodore's Cottage, was the prime mover of this lawn-tennis; the hospitable

dispenser of tea to the combatants, and the active
promoter of whatever healthful amusement gave pleasure to his junior officers, and to the few young people
on board the *Ascension.* The spirit of dissipation
seized upon us, and as a covering of cloud kindly promised to hide our folly from the contemptuous stars, we
threw off the "Sun's Parallax" for a night, and gave
ourselves up to mirth and revelry. The officers of the
Boxer being accomplished players on wind and
stringed instruments, we actually succeeded in getting
up a dance; a thing unheard of in the annals of the
island, and the few ladies did excellent duty to their
numerous partners.

Next day David returned to work, while I remained
behind, with the intention of joining a Garrison party
in a proposed excursion the following afternoon to
Wide-awake Fair, which was now at its noisiest.

Wide-awake Fair is a puzzling cognomen, and suggested to me, when I first heard it, a scene very unlike
the real one. "Fair" created a mental picture of
busy trafficking; puppet shows and penny-a-peeps;
"Sweetie-wives" and stalls of ginger-bread; dancing
bears and barrel-organs; lads with blue bonnets and
Sunday coats, and lassies "blythe and bonnie" carrying off their loads of "fairings" with gay laughter and
merry jesting. And "Wide-awake?" Perhaps the
trafficking without the jesting, and instead of North-
country lads and lasses, keen-eyed, swarthy Jews,
intent on their pound of flesh. But not on Ascension
could such transactions be possible, and the real

"Wide-awake Fair" turned out to be something very different.

"Wide-awake! Wide-awake!" is the response of thousands of baby birdies to the encouraging cry of their mothers—"Come here! Come here!" in the lessons of first flight. The noise they make is certainly "fair" like, hence the names "Wide-awake" (*Sterna fuliginosa*) and "Wide-awake Fair."

During our short and busy stay at Ascension, it was unfortunately not possible for us to study the habits of these birds. This I regret, the more that the popular stories about them vary, and, as far as I have been able to discover, natural history is provokingly silent on the subject.

Howard Saunders, F.L.S., F.Z.S., &c., in his book on "Sternæ" or "Terns" writes, regarding this particular species, shortly as follows :—

"It is said that at Ascension Island the 'Sooty Terns' or 'Wide-awakes' come every eight months to breed; *if true*, this is somewhat remarkable. The foot of this species is webbed to the extremity of the toes. The young are dark on the under parts."

But with regard to the time of their coming and going, the general opinion at Ascension is, that it is irregular and uncertain; that the birds always remain on the island until they have a young one ready to fly away with them, and no longer. Each pair look upon it as their duty to rear one child—each hen lays one egg; but if by any means *that* is destroyed, she lays another, and so on, till she is the happy mother of a chick. Some people think that, if the eggs were judiciously removed in some way, the birds would remain permanently on the island!   Certain it is, that when the great flood of 1876 swept away all the eggs from the Fair, the birds began to lay again, and they were never before known to remain so long on the island.

I have said that the Wide-awakes choose their nurseries for the most part among the rocks in the centre of the island.   The largest "Fair" which we now visited, lies between Gannet Hill and Riding-school Crater, about three miles from Garrison and two from Mars Bay.   Here David, attended by Graydon and Sam, met us, and so did the Wide-awakes with a noisy greeting.   Poor little things, how they shrieked in their excitement!   To say that there were thousands, conveys no idea of the vast multitude of birds that whirled around and above us—so close that one gentleman caught several, seizing them in his hands as they flew by.   One carried in its bill a tiny fish, which we took the liberty of examining, and, much to our surprise, found it to be no habitant of Ascension waters; so that this hungry little Wide-awake—about the size of an ordinary pigeon, only more slender and graceful in form—must have flown many a weary mile in search of its prey.   We restored to him his supper and his liberty.

Of course there was competition in egg gathering, at which I was singularly unsuccessful, feeling so confused by the deafening noise and so sickened by the

P

strong smell of guano, that my wits went a-wool-
gathering instead.

Moreover, I did not much appreciate the delicacy of
Wide-awakes' eggs, some of which had been gathered
for us a few days before by Tom,* our new marine
cook. These eggs, I had been told, were exactly like
plovers', but except in size and colour I could detect
little similarity. The white is certainly clearer and
more glutinous than that of a common hen's egg, and
the yolk is of a beautiful saffron or pumpkin colour,
such as I never saw in any other. We cooked them in
every conceivable way—in puddings, omelettes, pan-
cakes—fried, boiled, poached; and concluded that they
were most palatable when boiled hard and eaten cold.
In puddings I could not get them to " rise," but possibly that was owing to my bad management and no
fault of the egg. The thin shell is speckled very much
like that of the grouse, and is difficult to detect on
the bare stony ground on which the eggs are laid. It
is more by the excitement of the birds in the neighbourhood of their treasures, than by anything else, that
one discovers them; and so bold are they that the
female will hardly leave her post until actually thrust
aside.

It was very amusing on one occasion. Just as
we had scared a little hen from her solitary egg, her
lord and master swooped down to defend it, and stood
over the treasure screaming and flapping his wings in

* I think what follows of our story will explain my reason for
giving a fictitious name here.

a fury, and threatening to attack any one that dared approach him. I admired his courage so much that, had not this been my first find, I certainly would not have robbed him.

We gathered a good many dozens, but the eggs were by no means so numerous as I had been led to expect. I had been told that it was customary, on going to the Fair for plunder, to mark off and clear a space of ground, and then to sit down at some little distance and smoke a cigar till the birds should lay afresh. From the word "clear" I had conceived an absurd idea of the ground being so covered with eggs that it would require careful stepping not to crush them! This is decidedly not so, and, as I succeeded in finding only fifteen eggs, I should prefer to say that they are scarce. But "Honesty is the best policy," and I must confess that the St. Helena boys, who cater for the officers' mess and for the few private families in Garrison, sometimes carry off as many as *two hundred dozen* in a morning.*

The marines too, are large consumers, and we were struck with amazement at the pathetic anxiety of a stout lime-burner, who told us "I fear some'at must be wrong with me; last season I could eat as many as four dozen o' them Wide-awake eggs at a sittin'—but now I can only manage a matter o' two."

---

* I afterwards ascertained that we had visited the Fair before it came to its height; a month later the eggs were very much more numerous.

Poor plundered birdies! no wonder they shrieked.

Close by the Fair is to be found, in the shade of some high rocks, a tea equipage with an iron tripod on which to hang the kettle. It looked tempting, but we had no firewood, no tea, no water, and no time wherewith to take advantage of this considerate arrangement. Moreover, we were warned of the approach of night by the sudden disappearance of the sun behind Gannet Hill.

Just at the foot of the hill here there appeared to us the strange phenomenon of a brilliantly green flat, of about an acre in extent, looking like fresh spring-grass in the distance, and presenting a most striking contrast to the colourless barrenness everywhere around. On nearer inspection this oasis was found to be formed by a gigantic creeper (*Ipomæa maritima*), with large, bright green leaves, somewhat resembling those of the bay, but of a lighter shade, and bearing a purple convolvulus-like flower. One of the long tendrils, chosen at random, we followed up, and found it to measure seventy-two feet.

At Gannet Hill we parted from our friends: they going north to Garrison, and we southward to Mars Bay. We reached our tents shortly after sunset, to find Tom very cross at our having kept dinner waiting. Or perhaps it was my return that had up-set his precarious temper, for Tom could not be made to recognize female authority, and resented any fault-finding on my part, although to the master he was obedience itself.

But on board ship what else could one expect? There, " Woman's Rights " are unknown!

I had offended Tom too, very soon after his arrival, and I think a little soreness remained to the end.

One hot morning he had gathered, and presented to me, some dozens of Wide-awakes' eggs, with many impressive observations about the dust, and the drought of the journey. In a weak moment I rewarded him with a glass of grog—as much as I thought advisable ; but no doubt my ideas on the subject of grog are narrow, for the droughty marine felt insulted at the meagre draught and deliberately poured it upon the ground, with a cutting remark to the effect that, " Such a drop as that wouldn't wet his tongue."

I naturally felt indignant, and it was not until I began to study the comic side of my cook, that I found him interesting and forgave him. But it is wonderful how one is depressed by bad temper, even in the kitchen ; and Tom's surly greeting made me look back with regret on the bright welcome that Hill used to meet me with.

Poor Hill! he got no better, and had to be sent home by mail steamer. David went to see him in hospital before he sailed, and we both felt quite sad when his hammock and screaming " Polly " were carried off from Mars Bay.

Our community was so small, that each person and thing became an intimate friend and a part of our life.

Thus our servants occupied a larger place in our home
circle than they would nave done in England, and this
is my apology for writing of them so much. They
too, were separated from their fellows, which of course
was a much greater privation for them than for us.
We had a definite object, an absorbing interest, at
Mars Bay. We had numerous resources within ourselves, such as study, light-reading, and a large home
correspondence.   Moreover, we were *two*, and it is no
marriage if husband and wife are not each other's best
society.

Under these circumstances, we were naturally
anxious to include our servants in our life, and to keep
them amused as much as possible ; but now it became
more difficult than at first. Hill and Graydon had
been great allies, but the proverbial antipathy between marines and blue-jackets had no exception
in the case of Graydon and Tom. They did not
quarrel exactly, but they did not "chum up," as
our colonists say, and I fear I laid this to the charge
of Tom.

He was one of those " I told-you-so " characters,
never in the wrong, and with tender feelings. It was
these feelings I hurt, when I gave him too little grog ;
and the remotest hint of his having done anything
wrong was followed by a resigned expression of
countenance, and an air of injured innocence, which
was very ludicrous, after *my* feelings had ceased to be
aggravated by it.   For Tom was only one of a class—
one of those sensitive creatures that look sadly around

them, thinking how much better mankind would be if it could better appreciate them—making themselves the centre of their little world, the sun of their little system, and crying aloud if they are rudely pushed against and their centreship ignored.

CHAPTER XIX.

LAST DAYS AT MARS BAY.

Passing ships.—H.M.S. *Beacon.*—Graydon home-sick.—"Rover" bequeathed.—A new assistant.—"Melpomene."—Bad weather.—Disappointment.—Delightful orders.—Scarcity of lucifer matches.—Attempts in Taxidermy.—Our feathered friends.—*Sterna leucocapilla.*—Packing.—The Ways of *The Service.*—My ignorance.—The work is done.—Pulling down our altars.

SOME weeks now passed by pleasantly and busily, but so entirely devoid of incident, that I fear my readers would find in a daily chronicle of them only monotonous repetition.

Sometimes a pretty little donkey would peep over the rocks at us and scamper off again; sometimes a wild cat would mistake my larder for public property, and bring involuntary fasts into the camp. Almost daily a ship of some kind passed us; sometimes so far off as to seem a mere white speck on the horizon, at other times so close that we could easily read her signals without the help of a glass. "What news of the East?" "Is England at war?" "When is the mail due?" were invariable questions; and often, when outward-bound vessels found that a mail was expected soon, they would send letters ashore at Garrison, so that Ascension has still good claim to its old name, " The Sailor's Post Office."

About the middle of November H.M.S. *Beacon,* homeward-bound from China, put into harbour for coals and provisions, and her coming made another break in our little community, for at sight of her Graydon grew home-sick. Having exceeded his term of service on the West Coast by two years, he might have gone home with the *Orontes* in September; and David could not find it in his heart to oppose him now, much as we were likely to suffer by a change.

So Graydon applied to the Captain for relief, and was ordered to report himself on board the *Beacon* the following morning. Meantime another blue-jacket had been appointed to take his place at Mars Bay.

Poor Graydon! he was glad to go home, and yet we had quite a scene at parting. It was in a voice suspiciously husky that Rover was bequeathed to me, while David made a miserable attempt to whistle and look unconcerned when I begged him to decide whether or not I should accept the gift. I could not conveniently bring Rover home with me, and was afraid that Graydon might not like my having to leave him behind, but he was far too much excited to calculate for the future.

" I don't care where you leave him, ma'am. I am sure somebody will be good to the poor little chap, and I would like to leave him with *you*," said Graydon.

So Rover remained, and howled many days and

nights for his master, refusing to be comforted. It was very doleful ; but by-and-by the faithless doggie forgot, and became quite friendly and happy with " Captain," an English terrier which accompanied Brackley, Graydon's successor at the Observatory.

We had now reached the 21st November. Fine evenings up to the 9th had allowed David to finish what was absolutely necessary of the Mars triangulation, and since then hardly a star had been visible.

On the 12th he had intended to commence observations of " Melpomene," and by a similar process confirm the result of the Mars observations. Melpomene (the " Muse of Sadness ") is a tiny planet between 8th and 9th magnitude, and is almost lost in the multitude of minor planets that have been discovered within the last thirty years. But now she had been selected to help in the great work of fixing the sun's distance ; because it so happened that for some weeks about the 2nd of December, she would come a little nearer to the earth than the sun, and her small disc, undistinguishable from that of a star, would permit observations of extreme accuracy.

David had already completed a similar work, under less favourable circumstances, with the planet Juno at Mauritius in 1874, which, as a first experiment, had been entirely successful. It was with much interest, therefore, that we looked forward to the present occasion as likely to afford a determination of the Parallax, little inferior to that to be obtained by Mars. Not that Melpomene was so near to us as Mars—indeed, she was fully two and a-half times as distant—but the observations were likely to be two and a-half times as accurate, because the measurement of the distance between two minute points is capable of so much greater precision than that of the distance between a minute star and a large bright planet such as Mars.

The experience which my husband had had in the case of Juno led him to feel confidence in the method —a very satisfactory thing when the result of a year's labour in observation and calculation lies hid till the last step of the process. We therefore left Mars to retreat unmolested for a season, and turned in pursuit of Melpomene, who now, on the 21st of November, was threatening to escape us under cover of many cloudy nights.

Before the 2nd of December, the night of Opposition, a few good evening observations were obtained, but not a single complete morning set. Ever since the second week of October, the sky had invariably become overcast soon after midnight, and so remained until sunrise, rendering any evening observations for parallax useless for want of corresponding morning ones.

Well, we dared not grumble, so thankful were we for the fine nights of August and September. Every year, two or three of these minor planets would come into opposition under circumstances favourable for the purpose in question, but never again would Mars appear so glorious to our eyes on this earth; and, having secured the main, nay, sole object of the expedition, we could the more easily bear this comparatively trivial disappointment.

The conditions for the observation of Melpomene being more favourable before than after opposition, David was able to decide by the end of November that he would waste no more evenings in a work he could not perfect; but would rather make use of any clear weather that might occur, to determine yet more thoroughly the places of the Mars stars of comparison.

It was a fortunate decision. Every morning clouds covered the island, and sometimes heavy rain fell at intervals between 1 A.M. and sunrise. But this cloud and rain, which lost us the Opposition of Melpomene, gave rise to a very important announcement, made on the 4th of December. Brackley had gone to Garrison in the morning, to draw his pay, and on his return brought us the welcome news—" If you please, sir, the Captain's orders are, that from to-day we are on double allowance of water."

Delightful orders! but Tom of the rueful countenance at once reproved our enthusiasm by the correction, " Not *double* allowance, ma'am, only *full* allowance; we were on *half* before." This depressing logic my feminine mind could not follow, and I persisted in my rejoicings over two gallons of water in the place of one.

Speaking from a domestic point of view, the last days at Mars Bay were famous for full water-casks and a scarcity of lucifer matches! At the canteen the supply of these latter articles had become exhausted, and our stock on hand consisted of but two small boxes. These I guarded jealously, and David suffered great anxiety in getting his cigars alight, until, discovering a box of " Vesuvians " in his despatch box, he was by this treasure trove rendered independent.

The astronomical work of the last days at Mars Bay consisted chiefly in lunar and other observations for longitude and latitude ; and on the whole, the weather was propitious—much cloud and occasional showers, with opportune bands of clear sky intervening which served a good purpose.

My time was partly occupied in collecting, labelling and packing up specimens of our rocks, shells, sand, &c., in the hope of being able to know them better by subsequent study. We were also anxious to bring home with us some specimens of birds, but hitherto had always failed in our attempts to preserve them. We were without some of the chemicals necessary for the operation, and certain insects of Mars Bay seemed to revel in carbolic acid.

But fortunately, one well skilled in taxidermy came to our assistance at last, and, thanks to him, we were able to bring to England some of our Ascension feathered friends.

The delicate French-grey Boatswain Bird * (*Phaeton*

---

* Boatswain Bird seems to be a sort of general term used by the sailors to denote various sea-birds, but as far as I could gather, the Tropic Bird, or *Phaeton ætherius*, is the Boatswain Bird proper.

ætherius) with its graceful tail of twin feathers; the
broad-winged Frigate Bird (*Tachypetes Aquila*), which
I at first mistook for an albatross, and was much puzzled
by the gay scarlet pouch depending from its breast
This pouch, filled with salt water, serves the purpose
of a game bag, in which live fish are brought home by
the male parent to feed his dainty young, who evidently do not approve of "high" game nor tinned
salmon.

The Gannets (*Sula cyanops*) and Boobies (*Sula
leucogastra*) are well known, and are by no means so
handsome in shape or plumage as the Frigate and
Boatswain birds. Of the Wide-awakes, I have already
told all that I know, and only one other bird came
under our notice at Mars Bay. That happened in
this way.

One evening during our last week there, David was
sitting at the door of the tent at sunset, on murder
bent, and watching for a flight of Wide-awakes. I
am glad to say, it is against the game laws of Ascension to shoot these confiding birdies, but we only
wanted to secure one or two for scientific purposes,
and to this extent we had absolution.

We had heard that Wide-awakes grew grey with
years—that the young birds were of a uniformly sooty
brown, and only acquired their brilliant white breasts
with maturity. Now, having frequently noticed some
birds—altogether black, with the exception of a white
spot on the head—accompanying the Wide-awakes,
and closely resembling them in shape and size, we were

curious to know if these were really the same birds in a transition state of colour.

A right and left shot brought down a couple of the white-capped strangers, and then it was discovered that they were no Wide-awakes at all, although belonging to the same family. Mr. Unwin, the naturalist, who had so kindly helped us before, pronounced them to be " *Sterna leucocapilla*," a very rare species of *Tern*, hitherto unknown to Ascension.

Of these Mr. Saunders says, " This form is apparently less widely diffused than some of its congeners. Mr. Gould's specimens were obtained at Raine's Islet, Australia, where the bird is said to be very abundant. There is a specimen in the British Museum, from Bristow Island, south coast of New Guinea; and the United States Exploring Expedition found it breeding at Panmotu Island, where its single egg was deposited upon the bare ground, instead of in a nest. There is no grey about the head or cheeks, but, with the exception of the white crown, the whole plumage is of a sooty brown."

Australia! New Guinea! Panmotu Island! How came the little wanderers so far from their homes? For up to this time Ascension had been no home nor colony of theirs. They had never been known to breed in any part of it, and on no occasion did we observe them to fly inland. They invariably steered their course in one direction—against the wind, and in their flight just skirted our shore.

Why? And wherefore this sudden appearance in

such numbers at Ascension? Did some ocean rock
which had been their home, sink suddenly beneath the
waves, leaving no rest for the sole of their foot, until a
new ark was found? Let wise men answer. Verily the
birds of the air have many things to tell us, undreamt
of in our philosophy.

Excepting Guinea Fowl and the universal Dorking
with variations, the only land bird that we saw on Ascension was a pretty little finch called the Averdavat
(*Estrelda amaudava*), ringed round the body with two
shades of grey, and having a red bill and red spots on
the breast and under the eyes. It is much valued
there, and also in this country, as a cage bird. Indeed,
almost the first drawing-room which we entered on
returning to London, contained a cage in which were,
hopping from perch to perch, two little twitterers that
looked strangely familiar, and with the glad feeling
of having met with old friends, we exclaimed
" Averdavats ! "

After the packing of my " clinker " box was satisfactorily accomplished, I set about collecting our
household gear, which certainly was not improved by
its sojourn in tents. Tom was not so efficient an
assistant in such work as Hill had been. He was
always bringing forward some remarkable method for
doing everything better than anybody else, and
although the method generally failed, his confidence in
himself was never shaken thereby, because it was
always somebody else that had caused the failure.
Wonderful indeed were his achievements in his own

imagination; and the way he patronized me, and pardoned all my mistakes because of my ignorance of the Service, was most amusing.

"Ah! you see you don't know the Service as I know it," was the refrain of all Tom's excuses for me, and I certainly doubted my ability ever to know the Service according to Tom. It was so complicated and had so many peculiar traits of character.

I was constantly having such shocks to my memory about this time as,

"How many cups and saucers had you, ma'am?"

"Four," I answer at a venture.

"Well, there are only three now, and one don't count for much, for the saucer is chipped, and the cup leaks ever so."

Or again, "Please, ma'am, how many glasses came out?"

"Six," and this time I am sure.

"Just one now, ma'am."

"That's very awkward," I answer; "how are we to replace them?"

"Oh!" says Tom, "that don't matter, I've kept the pieces, which will count just the same as if the glasses were whole. It is the way of the Service—so be that you return the articles, ma'am, it don't matter *how*."

Notwithstanding these wonderfully accommodating ways of "The Service," I still doubted the propriety of presenting the Admiralty with broken glass, and persisted in expressing my intention of replacing the damaged goods, if possible. Whereupon Tom would look upon me with a pitying smile, as he repeated his favourite formula, " If you knew the Service as I know it, you wouldn't think of such a thing, ma'am."

I thought such knowledge must be difficult to acquire, so gave up the attempt, and turned my mind to matters more fitted to my understanding.

We wished, if possible, to have our " flitting " over before Christmas Day, and were most fortunate in being able to manage this. On the 23rd, the last moon occultation for December took place, and as the night was clear David was able, not only to secure this, his last chance for longitude, but also to complete his determination of latitude by observing the transits of stars in the prime vertical.

Then the work was done ; that is, of course, apart from the laborious calculations which must ensue. As I write now, these are still unfinished, but the reductions are sufficiently advanced for me to say, almost with certainty, that our six months of anxiety have been crowned with success.

How glad we were when the Observatory dome was shut for the last time ! So grateful too, and ready thoroughly to enjoy a merry Christmas. Early on the 24th, the turtle-boat came for our household goods (the Heliometer being left meanwhile), and our tent life was at an end.

Thankful as I was to have the work finished, yet I had struck some roots in the clinker which could not be pulled up without a wrench, and it was with a lump in my throat and dim eyes that I said good-bye to Mars Bay.

Those are enviable people who retain in larger growth a child-like satisfaction in the novelty of coming and going—who are glad to seek new places, and as glad to leave them behind, thereby ensuring for themselves two pleasures in place of two trials, such as afflict the heart that is wedded to old scenes and old faces. To me, a new home, however pretty, gives no rest until I get to know it; then I love it, no matter how plain its features. Then it becomes a friend who has shared my joys and sorrows, seen my cares, and carried in its bosom my most sacred treasures. It makes my heart ache when I must leave this well-beloved friend behind, and memory counts with a sigh one more of life's regrets. My life has been so ordered, that I have loved and left many homes; but I hardly looked upon Mars Bay as one, until I saw it for the last time. Then I knew that I had set up my household gods in those tents as surely as in a costly temple, and that this was another pulling down of their altars.

CHAPTER XX.

CHRISTMAS IN GARRISON.

Five precious Books.—Christmas without Holly.—Roast Beef and Plum
Pudding.—A Musical Entertainment.—The Tom-tom.—A motley
crew.—The Master of Ceremonies.—No cook and no dinner.—
Eggs and bacon.—The Captain's Office.—Exit Tom.

CHRISTMAS morning found us again at Commodore's
Cottage, with everything around looking much the
same as during our first occupation. Only now there
was no Observatory on the croquet lawn, while, lying
snugly in a strong box were five books of manuscript,
containing " Observations of the Opposition of Mars
at Ascension, A.D. 1877." These differences contributed much to our comfort, and even with the
thermometer at 89° F. and not a bit of holly in
the land, I was prepared to enjoy my first summer
Christmas.

Having fallen asleep with the sound of imaginary
"waits" and "Christmas Carols" ringing in my ears,
I was awakened at sunrise by a very different sound—
the beat of tom-toms, accompanied at intervals by a
sort of hoarse monotonous chant, which we were given
to understand was intended as a song of rejoicing on
the part of the Kroomen.

As we walked down to church in the bright morning, our little Garrison looked quite gay and festal. Flags of various colours and devices were flying over the different mess-rooms, and all the men showed clean and trim in holiday attire. Alas! I fear some of them were less clean and trim before the shades of night had fallen; but I must not make the behaviour of a faulty few, a type of the whole. Generally speaking, the men enjoyed their Christmas rationally, and not a single disturbance annoyed us in Garrison during their three days' idleness.

Our Christmas Day service was very short: advisable, no doubt, on account of the heat, which made it difficult to keep the attention fixed or the mind vigorous for any length of time. After service, and in the heat of the noon-day sun, the men dined on the traditional roast beef and plum-pudding with "trimmings."

I wonder how hot it must be before an Englishman would give up his heavy Christmas dinner? Apparently the temperature of Ascension made no difference to his enjoyment of it, although *we* could not speak on this matter from experience. But thereby hangs a tale presently to be unfolded.

When we came from church, Tom begged permission to dine with his messmates, which I readily granted him, on condition that he would return at four o'clock to prepare our dinner, for we had declined many kind offers of hospitality for the day. Then Sam, in his turn, longed to help his countrymen at making

holiday, so we were left to spend the day with deserted
kitchen. Sam was delighted to be off, and told me
with much glee that "they going to have great fun
down at Krootown," which 1 suppose they had, for the
sound of the tom-tom, accompanied by yells, was borne
on the wind all the afternoon, and ultimately we had a
special benefit of this unmelodious music.

As the sun went down the horrid din waxed louder;
the distance, which had lent its only enchantment,
gradually decreased, until at last our verandah was
surrounded by a crowd of Kroomen, dressed in the
most extraordinary variety of costumes, yelling, beating the tom-tom, and dancing what they pleased to
term a war-dance. When the troop "hove in sight,"
I was sitting alone on the verandah, trying to catch
the first cool breath of the evening breeze, and my
nerves were hardly strung to a pitch sufficient to enjoy
this musical entertainment in full, so I hastily re-
treated within, preferring to bear it in the privacy of
my chamber. This was not according to the programme, however. An insinuating knock from Sam
came to intimate that I was expected to go outside.
"They come wish you happy Christmas, ma," he
called out; and if noise has anything to do with good
wishes, they were certainly expressed with a will.

"Wah-y-a-wah-wah! Wah-y-a-wah!" in every
octave from the shrillest soprano to the deepest bass,
greeted me outside, and one little fellow, black as
Erebus, was seated on the verandah steps, playing
with all his might on their favourite tom-tom. This is

a very primitive instrument of simplest construction, and seemed to me to consist of a small barrel with an end knocked out—across the opening a dressed hide was stretched, and this was beaten with the palms of both hands.

The poor things must have been very tired and hot with all this dancing and yelling and beating, to say nothing of the unaccustomed amount of clothing they wore. One dark figure was crowned with the orthodox "tall" hat—the abomination of civilization—and it evidently impressed the wearer with a crushing sense of his importance. Another woolly head was encased in a white muslin "Dolly Varden;" while a pair of large blue spectacles adorned the nose of a third. The whole had the most comical effect, and they all grinned and grimaced and laughed so, that it was quite impossible to be in the least impressed by a great tin sword, which was evidently intended to furnish the warlike element in the performance.

By-and-by David appeared, and then the shouting grew more tumultuous; partly from good will, let us suppose, but I fear that a new hope of bubbly-water was what chiefly increased the excitement. After we thought that they had made noise enough, we hinted as much, and (saying to his conscience, " Christmas comes but once a year") David produced a bottle of rum, which was promptly taken in charge by the head Krooman, " Chop-Dollar." Dressed in his best blue jacket and white duck trowsers, he seemed to be acting as Master of Ceremonies, and wore an expression of

face as solemn as if he were conducting a funeral instead of a holiday frolic.

After the excitement of the war-dance was over, I
began to grow anxious about Tom and dinner, and on
peeping into the kitchen, what was my dismay to find
no Tom—no fire—no beef—no plum-pudding ! However, Sam, who had now returned to his duties, comforted me with the information that Tom had carried
off our dinner to be cooked in the marines' galley ; and
I took no further care about the matter until six
o'clock, when the pangs of hunger again urged me to
action. Still no sign of cook or dinner, and scouts
were now sent out in all directions in search of the
" missing," but with no result. Ultimately David
went, and his finding of the case was—Tom drunk,
galley locked, fire out, and nobody knows nothing
about nothing !

Here was a nice state of things on Christmas Day :
no dinner, and nothing in the larder ; for in these
climes one cannot be fore-stocked in fresh provisions
at Christmas time as in frosty England. The best
result of my forage was eggs and bacon. It was very
hard, and I felt very cross ; while, to make the matter
worse, David was so unmanly as to treat the misfortune as a good joke, except in so far as Tom's conduct
was concerned. *That* he felt he must punish as the
same fault had occurred several times before ; and
only the previous evening, Tom had been warned that
the next offence of the kind would be reported.

There is surely no duty in life more disagreeable

than that of fault-finding, and the necessity for it in
the present case was an unpleasant thought on Christmas Day ; but after making up our minds, the flavour
of it vanished from our bacon and eggs, and we contrived to spend a delightful evening. A cup of tea
smoothed my ruffled humour, and we congratulated
ourselves on the probability of a more joyful awakening next morning, than if we had dined heavily à la
John Bull.

I certainly never expected to see our Christmas
joint again, thinking its ashes had gone to swell the
great cinder heap on which we lived. But, amazing to
relate, it appeared next morning at breakfast—cold and
none the worse ; so did our cook—also cold but not in
so good condition, poor fellow ! David *almost* changed
his mind and would fain have taken no notice, especially, if the truth must be told, because reporting
Tom was tantamount to dismissing him, and we must
thus punish ourselves by being short-handed during
the busy days before leaving.

But remembering the danger of an unfulfilled warning, he stuck to his word, and Tom was taken to the
Captain's office—had thirty days' grog stopped, and
on my husband's relieving him from duty with us, he
was sent to tend the bullocks at the Mountain.

Let us hope he was more active in ministering to
their physical needs than he had been to ours. Exit
Tom !

CHAPTER XXI.

ABOUT THE KROOMEN.

THE day after Christmas was also a holiday in
Garrison, and we had a visit during the morning from
our old friend Sam the first, in Sunday suit, along
with another Krooman whom he introduced as his
" chum."

"We come wish you merry Christmas-time," said
Sam ; and we thanked them, hoping in turn that they
had had a happy Christmas.

"No, ma, me no happy Christmas, other man drink
it all," said Sam.

I didn't quite understand him at first, but it gradually dawned upon me that, to poor Sam's thinking,
Christmas happiness was in proportion to the amount
of bubbly-water he could consume.

How I wish we could show our good will to these
poor fellows in some other way than by giving them

" something to drink ! " But there are so few things that they appreciate. Some of them, indeed, accept money eagerly, but for the most part they " no care." And it is no wonder, for all the money they earn has to be given up to the Big Brother.

Although it is an undoubted fact that the Krooman cannot live as a slave, and has been known in slavery to starve himself to death, yet this Big Brother system almost amounts to bondage. It has its origin in this way :—

Very often a well-to-do, long-headed Krooman seeks out a few miserable starving countrymen in the interior, brings them down to Sierra Leone, feeds and clothes them there for a time, and then ships them on board a man-of-war, on condition *that they bring him back all the money they earn.*

Each of our ships of war on the West Coast of Africa is allowed to employ a fixed number of these men (proportional to her crew), and they are invaluable in that climate for all hard work involving exposure to the sun. They make splendid boatmen, and are able to maintain communication with the land when no blue-jacket could take a boat through the surf; so the Little Brothers, paid by our navy beyond their wildest expectations (9*d.* a day), soon begin to discover that they have made a bad bargain. Yet their sense of honesty, combined with their fear of Fetish, is sufficiently strong to make them keep to it in the letter, though they do not scruple to break it in the spirit. Either they grow reckless about their savings,

or else spend every sixpence they earn in goods to
which the Big Brother can lay no claim.

Clothing is their favourite mode of investment, and
some of their wardrobes must be a rare sight.   There
is nothing they won't buy, especially if it be European ;
and no doubt they astonish their friends at home by
donning their heterogeneous attire when they return
to Fatherland.   Doubtless also they deck out their
mothers and wives and sisters in serge gowns and
Dolly Varden hats.   Filial affection is one of the
many virtues of this African race, and they store up
the most extraordinary treasures for their old people.

Some time before our arrival there was a sale by
auction of the wardrobe of an officer lately deceased,
at which the Kroomen bought largely ; and as they
bore off their various purchases on their persons, some
striking figures were presented to the sketcher.

" Chop-Dollar," who is a reputed Crœsus, became
very much excited, bid wildly for an Ulster, and having
secured it, proceeded at once to put it on.   Then a
dress-coat fell to his share, and that was put atop of
the Ulster, tied by the arms round the neck.   Next a
tall hat and hat-box.   The hat was promptly clapped
upon his head ; the box he placed between his feet,
and continued bidding until it was filled with handkerchiefs, collars, ties, and other small goods.   Then,
thinking no doubt that he had done his duty, and de-
frauded the Big Brother of enough of money for a
time, he proceeded triumphantly to Krootown, bearing
the loaded hat-box, Krooman-like, on his head atop of

the hat, and accompanied by a dusky crowd, grotesque
in newly-purchased attire, and all laughing and chattering in their usual good-humoured way.

We became deeply interested in the history and
character of these men—whose industry, honesty, and
imperturbable good-nature make them such valuable
servants. Physically, the Kroomen are well-formed; of
a medium height, and stoutly built, with woolly pates,
and of an open, pleasant countenance, black—very
black, though it be. A stripe of blue tattooing runs
down from where the wool begins to grow, to the point
of the broad flat nose. The mouth is better cut, and
the lips less thick than those of the real African negro ;
but, on the other hand, they cannot boast of their
beautiful teeth, and the Kroomen further disfigure
theirs by filing out a triangular space between the two
front ones. Yet the smile is pleasing, and has a
wonderful brightness in it, lighting up the whole of
the dark face like a sudden sunbeam, and a kind word
has the power of calling it forth at all times.

The moral nature of the Krooman is undoubtedly
high, and one eminently fitted to receive Christianity.
Many of them indeed do attend the English Church
service, and a few I believe have been baptized ; but,
generally speaking, the reverence that the greater part
of them have for Christianity is due to a belief that
" white man's Fetish better than black man's."

This Fetish worship is common to all negro races,
and exists among them in many other parts of
the world under slightly different forms. Kingsley

found it very strong in the West Indies, and
gives much interesting information on the subject;
but the system seems to have become so involved that
it is impossible to trace it, the worshippers themselves
being the most ignorant of what they worship.

With regard to it Kingsley says :—" Here, perhaps,
I may be allowed to tell what I know about this curious
question of Obeah or Fetish worship. It appears to
me, on closer examination, that it is not a worship of
natural objects ; not a primæval worship; scarcely a
worship at all; but simply a system of incantation
carried on by a priesthood, or rather a sorcerer class ;
and this being the case, it seems to me unfortunate
that the term 'Fetish worship' should have been
adopted by so many learned men as the general name
for the supposed primæval Nature-worship. The
negro does not, as the primæval man is supposed to
have done, regard as divine (and therefore as Fetish
or Obeah) any object which excites his admiration;
anything peculiarly beautiful, noble, or powerful ; anything even which causes curiosity or fear. In fact, a
Fetish is no natural object at all; it is a spirit, an
Obeah, Jumby, Duppy, like the 'Duvvels' or spirits
of the air, which are the only deities of which our
gipsies have a conception left. That spirit belongs
to the Obeah, or Fetish-man, and he puts it by magic
ceremonies into any object which he chooses. Thus
anything may become Obeah as far as I have ascertained. In a case which happened very lately, an
Obeah-man came into the country, put the Obeah into

a fresh monkey's jaw-bone, and made the people offer to it fowls and plantains, which, of course, he himself ate. Such is Obeah now, and such it was when the Portuguese first met with it on the African coast four hundred years ago."

As far as I can gather, the Kroomen believe strongly too in the power of certain charms, called "gre-gre," to propitiate Fetish. These charms are generally worn on their persons, and may consist of a finger-nail; of a lock of hair, human or belonging to some animal; of a couple of small pebbles carried in a bag round the neck; indeed, of *anything*, as Kingsley says of Fetish.

But at the same time one must not confound the two. The gre-gre is merely a charm used against the power of Fetish, which I never knew to be associated with anything beautiful in nature or art, anything calculated to inspire feelings of awe or admiration—but rather of horror and disgust. Indeed the principle of Fetish-worship is not love or even reverence, but fear. The Obeah is always an evil spirit to be propitiated, and his trembling worshippers enjoy and envy the boldness of those that dare to defy him.

A naval officer told me that at Sierra Leone he once bought an unshapely block of wood (with a rude head carved on one end of it), which was worshipped as containing the spirit of Fetish. This thing he set suddenly upon the deck one evening when the Kroomen were assembled; and then contemptuously kicked it over, wishing to try what effect such an act of sacrilege would have upon them. For a moment the black

faces looked aghast—almost pale, and then each ex-
panded into a broad grin.

"Will your Fetish hurt me for this?" asked their
Captain.

"Oh, no; Fetish no can touch white man."

Poor souls! it is no wonder that they long to be
white and beyond the power of this evil spirit.
Indeed, the great desire of their lives is for fair faces,
and the photographer at Ascension, who is much
patronized by the Kroomen, has frequently been
offered double price for an over-exposed picture,
because "it make them look white!" Perhaps they
thought it would be a gre-gre, and defy the power of
Fetish.

A considerable traffic goes on in these charms.
An unlucky man begins to get angry with his gre-
gre, and looks with envy on that belonging to a
more fortunate neighbour, who may, in his turn, be
casting longing eyes on some other gre-gre which he
wishes to purchase, while he is by no means unwilling
to part with his own for a fair sum.

While we were at Ascension a favourite pipe disappeared from Krootown, and after some time the en-
raged owner recognized it in the mouth of a comrade.

"That my pipe." "No, *my* pipe," persisted the
thief, who refused, with the most violent protests, to
give it up.

"Very well; Fetish will have you before four o'clock."
But the guilty Krooman was not afraid; he laughed and
said, "Me no care—my gre-gre better than yours."

Now, strange to say, that same afternoon about three o'clock, when the Krooman who had stolen the pipe was throwing away the refuse of the turtle that had just been slain, by some accident he slipped from the pier-head into the sea. These Kroomen are essentially an amphibious race, and seem to feel as much at home in the water as elsewhere, so that no great harm was done; but the cold bath stirred up a guilty conscience.

With haste and dripping clothes, the affrighted Krooman ran to restore the pipe, and at the same time to offer all his possessions for the powerful gre-gre (a bit of hair in a dirty little bag), that had brought upon him such speedy retribution. But no; its present possessor would on no account part with it, and so lose the pre-eminence it gave him over his fellows; for ever since this occurrence he has been much "esteemed and respected," and I have no doubt will die " rich and deeply regretted."

Well, it is not very long ago since our dear, blundering old Scotland believed in gre-gre, called by another name; and all over the civilized world at this day, methinks, men honour men less for their manhood than for their possession of the magic " gre-gre," called in our tongue " success."

After some questioning I found that my Sam was no believer in Fetish, and he laughed when I asked whether he wore a gre-gre. " No, ma, that no sense."

"Then, are you a Christian, Sam?" I asked; for he attended our English service regularly.

"No, ma," he replied, with a shake of the head, " but me want to believe what you believe—you tell me how."

Poor Sam! I felt that he was asking me for bread, and I could only give him a stone; for this conversation took place when our days at Ascension were growing few, and I had the bitterness of knowing that, in my anxiety to help the mind, I had left the spirit unaided. " Time " for my opportunity was almost gone, and " Too late " was near at hand; but I did the little I could, and I had a willing pupil.

By this time he could read at sight with tolerable ease, and before I left, he read to me the 14th chapter of St. John with evident understanding and without a mistake. A little New Testament was my Christmas gift to him, and he promised to persevere in the study of it.

Has he done so? I hope for the best, but my heart was heavy when I parted from him, and is so now when I think of time wasted and a precious opportunity lost.

CHAPTER XXII.

CLINKER CEMETERIES.

"In boxes."—Waiting for the Mail.—Dead Man's Beach.- -Garrison
  Cemetery.—The music of the waves.—A coral strand.—The
  "Blow-hole."—The Rollers again.—The Heliometer in danger.—
  Its fortunate escape.—Volcanic scenery.—Comfortless Cove.—
  Lonely graves.—A vision of the past.

BUT the Kroomen have betrayed me into a long
digression, and it is now absolutely necessary that I
return to Commodore's Cottage and its inmates.

We had hoped to spend our Christmas holidays at
Green Mountain, but having no authentic information
as to when the Mail would arrive, we dared not venture
far from harbour. Counting from the departure of the
previous Mail, she was due on the 10th of January;
but the captain of a barque which arrived from the
Cape on Christmas Eve, brought word from the agents
there, that the steamer calling next at Ascension
would leave Cape Town on Christmas Day. In this
case we must expect her on the 3rd or 4th of January.
It was very annoying to be "in boxes," and in a state
of uncertainty about our departure for ten days, but it
would have been infinitely more annoying to be left
behind. Accordingly, we gave up the idea of removing

our household to the Mountain, and contented ourselves
with making short excursions from Garrison instead.

It was now Ascension mid-summer, and there was
light enough for a good walk after five o'clock. Our
first spare afternoon was devoted to "Dead Man's
Beach," which I have already described as lying south
of Pierhead, and which, notwithstanding its gloomy
name, is bright and life-like, as the blue waves dance
in the sunlight, and break in quick succession on the
glistening sand. But about 100 yards from the sea,
where this pretty white sand runs into the black
clinker, the name is justified, for here lies the
Garrison Cemetery, very full of graves and very
dismal.

Not that the one condition is the necessary consequence of the other. It was not the graves that made
it dismal, but the crumbling headstones covered with
black dust; the wall, broken down in many parts by
the last heavy rains, half-burying some gravestones in
its ruins; one or two open graves, and tools lying
about for making others; the perfect barrenness everywhere, for not the tiniest flower bloomed within or
without. It was indeed a picture of death and decay,
and the sea sang a constant dirge for the lost lives of
the many brave sailors that lie buried here, within
sound of her mourning voice.

After leaving the Cemetery, we continued our walk
close to the water's edge, enjoying the fresh breeze,
and looking out for a curious beach phenomenon
somewhere about, called "The Blow-hole." Not know-

ing exactly what to expect, I fancied that a disturbed pool, which we now chanced upon, was the object of our search. Here the waves were surging through an underground passage, sending forth a cloud of spray from among the rocks, accompanied by a rich, low musical sound. The music had a metallic character, and something very like it we had often heard and wondered about at Mars Bay. There, it was audible only at certain spots and at certain times. In the Transit Hut it was nearly always to be heard, but when the rollers were in, we could hear it also in the Heliometer House, and in the open air.

We often puzzled over it, and concluded that it must be caused by the percussion of the waves against the rocks, which, in many cases, were of such a nature as to ring like a piece of metal when struck.

While we stood watching the waves and listening to their music, one of the hospital patients (who was strolling about idly, poor man, with a broken arm) came up to us, and, in answer to our inquiries, he told us that the Blow-hole was still some little distance off.

Ten minutes' walking brought us to it. At the extreme end of the sandy beach, or more poetically and more truthfully speaking—the "coral strand," we found a curious natural fountain playing.

"Not a fountain," said my husband, "but a champagne bottle uncorked!"

From a little hole, not a foot in diameter, in the

flat-surfaced rock, a stream of water suddenly gushed
forth; and, rushing up twenty or thirty feet, broke
into spray; then vanished as suddenly as it had
appeared, leaving the narrow vent quiet till the next
surge of the waves, when again the fountain leapt
forth like a thing of life and sudden death. It was
pleasant to watch it—with its fitful activity, so unlike
the steady, constant motion of the waves, which were
coming and going with the dull monotonous sound
that makes one sad or sleepy according to the mood.

To-day there was none of the sharp crash that the
rollers bring with them—the bay was calm, and we
regretted it, because the Blow-hole takes much additional life from the rollers. Those inconvenient, dis-
obliging, unaccommodating rollers! They declined to
satisfy our desire for a "spectacle" this afternoon, but
next day they created one, undesired, and almost
caused us a terrible misfortune.

With a view to being prepared for the arrival of the
Mail on the 3rd of January, my husband returned to
Mars Bay on the 27th of December, along with a party
of blue-jackets and marines, intending to dismount the
instruments and bring them back with him in the
steam-launch, as he had now more confidence in the
safety of the landing. Meantime, a lighter was
moored in the bay, waiting to be loaded once more
with our baggage.

When they got to Mars Bay at 9 A.M., rollers were
threatening, and my husband thought it advisable to
send off the most precious and delicate part of his

cargo first. So the Heliometer-tube was quickly dis-
mounted, fitted carefully into its case and carried by
four men down to the beach, where it was placed on
the seats of a dingey to be sculled out to the lighter.

But this dingey was aged and frail; indeed so often
had it been repaired, that current gossip reported a
structure of tea-lead and brown paper to have taken
the place of the original dingey. Be this as it may,
the little boat was unfortunately caught in a Scylla
and Charybdis condition by a mischievous roller,
and cast on the top of a rock in mid-channel. The
men got her off almost immediately, and the shock was
not a severe one, but unhappily severe enough to knock
a hole in the bottom of the fragile dingey, which at
once began to fill. They sculled out with all haste,
but by the time they got along-side the lighter, the
boat was filled nearly to the thwarts, and the Heliometer was saved from damp only by having been
placed upon the seats. By dint of baling (the dingey
being lightened of the Heliometer, which was now
safely on board the lighter), they got her back to
shore, but she gave no hope of being of any further
use in taking off the other things.

Then, after seeing everything dismantled and carried
to the beach, David set off to Garrison to beg for
another boat. He arrived shortly after noon very hot
and tired, and his unexpected appearance gave me
quite a shock. Are the difficulties of the Mars expedition never to be at an end? I thought. But how
thankful I felt not to have *seen* the Heliometer being

sculled out to sea in a sinking boat. The excitement
would have been more intense than pleasurable. And
how I wished the "uncanny" instrument safe at
home!

It seemed as if Mars Bay claimed a right to it, and
resented our carrying off her *raison d'être*. But we
could not afford to leave such a precious souvenir; and
the same evening another dingey took everything on
board the lighter, which was at once towed round to
the Pierhead, there to await the arrival of the Mail.

For the next few days my husband was busy packing. I too, had some of that work to do on a smaller
scale, and there being no one to cook for us except
a blue-jacket, who was strange to the art, the days
at Commodore's Cottage were hot and fatiguing this
Christmas-tide.

We always found time and inclination, however, for
our evening walk, and one day some friends joined us
in an excursion to Comfortless Cove. The distance
from Garrison is not great—about two miles, I should
think; but all Ascension miles must be multiplied at
least by three, in order to reduce them to the fatigue
unit of the English mile; and, despite the assistance
given by my never-failing friend Jimmy Chivas, some
of us found the way long and tiring. One or two
other mules were at our disposal; but unfortunately
the *Ascension* had only one side-saddle on board, so
three ladies had to make alternate use of Jimmy, and
trust to their alpenstocks for the rest.

This time our road lay in exactly the opposite direction to that we had taken in seeking for the Blow-hole. We now turned our steps towards Long Beach, past the turtle ponds. At first we skirt the foot of Cross Hill, whose steep sides of red cinders keep off all view and all coolness; but, these past, the breeze again rushes down upon us, the country opens up, the " Three Sisters " rise gracefully from the plains, and away in the distance " Green Mountain " and its floating clouds fill the east horizon with a beautiful mystery.

Beyond Long Beach, a point of rocky ground runs out into the sea, very much like South Point, and this we had to cross in order to reach Comfortless Cove, in the same way as South Point is crossed in going from Mars Bay to Gannet Bay. But here the rock is more disintegrated, easier to get over, and less picturesque. The white colour is mostly absent, and the prevailing tint is a deep red, changing through purplish slate into brown.

It was here ugly, dusty, and temper-trying; but looking up through the quiet hills, a sense of beauty was borne upon one unknowingly—that beauty which seems to belong so entirely to volcanic scenery—the beauty of calm after storm, of peace after tumult; the beauty of the seamed, fire-furrowed face of a quiet crater, which, like the beauty of wrinkled faces that have passed through the toil and fire of life, demands from the heart a greater tribute than mere admiration, and gives it in return a feeling of rest and thankfulness that the hot fight is over.

After passing over this dusty point, where the sea was for the most part hidden by the high rocks that edge the shore, it was a pleasant surprise to catch all at once the blue gleam of the water running into a tiny beach of white sand, which narrowed into the gully that had appeared, in the distance, but as a crack in our rough road. On the other side of this gully the ground rises perpendicularly, and forms a table-land, on which paths have been traced and broad level spaces cleared of ashes and clinker. Altogether, the place presented to us just such an appearance as, no doubt, Mars Bay presents, now that our tents are gone, to any stranger wandering on the southern shore of Ascension.

But here there was something besides. As we looked over the edge of the sea-sawn gorge, we saw a little cluster of white-washed graves lying in its bosom. Here there were no broken, crumbling walls; for the island shore had thrown itself around the sacred spot, sheltering it in faithful arms, strong and sure as its own existence; and the sleepless sea kept watch.

We did not go down into the valley, but rested on its rocky side; and here a waking dream stole over me, born of the sad scene and of the words of our guide—" A ship in quarantine for yellow fever landed her sick here, and many of them died."

Strong men are busy pitching tents in nervous haste, for wives and comrades are sickening in the sun, and there is none to help—no friendly neighbour to offer a cup of cold water to parched lips, no kindly hand to smooth a fever-tossed pillow. They are alone with God and with their sorrow, and some of them are sick unto death.

Now I can see a sad company of men, bearing living, dying burdens up the steep shores from where the plague-stricken ship lies anchored; but not a sound breaks the stillness, save an occasional moan which the toilers are too sad to answer, for *they* are bearing the future—the heaviest burden of the human soul— perhaps are envying those whose sufferings are passing away, and fearing, " Shall there be no man left to bury *us ?* "

Again and again I see sad, silent processions wending their way down into this sheltered nook, growing smaller, more sad and more silent each time that another and another member of the doomed company is borne to his last resting-place, until at last my eyes grow so dim with tears, that past and present are blotted from my sight.

How many died I know not, but it seemed to me as if all must have suffered equally—the survivors as much as those that are left behind in the little valley— struggling with Death in this dreary solitude, living in close communication with him, watching his ravages and waiting for his coming. Did any go mad, or did Christian faith and the courage of noble souls soften this awful experience into a gentle memory, refining the heart? God knoweth and God judgeth. " Let the earth rejoice."

CHAPTER XXIII.

CHRISTMAS HOLIDAYS.

NEW YEAR'S DAY, 1878, was a hot day in Ascension, and we tried hard to keep cool by recalling former New Year's Days spent in Scotland, much to the disadvantage of the present one.

What a burden life becomes when its chief end is to war against heat! Life, did I say? It is only existence in such latitudes, and with brain half-awake you speculate dreamily about life, with its hurry and feverish bustle, as a thing far off and beyond you ; and if sometimes you try to grasp it, nerves and spirit fail, you miss it, and, worn out with the effort, sink back into a deeper lethargy than before. That is to say, if you do not wear some gre-gre strong enough to defy the evil power of indolence—a Fetish too evil and too powerful in these climes to be easily overcome. But

English pluck and Scotch endurance can do it. Stay at home, or hang these gre-gres round your neck.

In Ascension each man wore one; and at six o'clock this New Year's morning my husband and two of the island officers were hard at work, practising for a rifle-match that they were to shoot the same afternoon against three officers of H.M.S. *Sea-gull*, then in harbour.

David was sadly out of practice; neither did the scoring of his allies, in this preliminary canter, give much promise of success, especially as their opponents had a high reputation as marksmen. The fear that the island should be beaten by the strangers, was strong enough to get up an excitement among us positively alarming in such weather; and the spirit of " buckling on the armour " showed itself to be still alive in wifely bosoms.

At four P.M., the tourney commenced, and from the door of Commodore's Cottage I had full view of the range. I could see the marker's flags as they rose and fell—yellow, outer—blue, centre—red and white, " bull's-eye; " but this became monotonous, when I could not see who fired the shots. So by-and-by two other ladies and I walked across to a tent, that had been placed near the range for onlookers. There we watched each shot with great interest, but the buckling-on-the-armour spirit, as well as every other sentiment of the fine old times of chivalry and romance, fled at sight of white flannel suits and braided uniforms covered from top to toe with the Gregory's-powder-

coloured dust of the country. What rueful figures our
knights presented! They shot very badly too, but
fortunately the " Seagulls " shot worse, and H.M.S.
Ascension came off victorious by nineteen points.
Bravo, the *Ascension!*

On the day following I had another excitement, more
peculiarly my own. With a *ménage* consisting of a
blue-jacket and a Krooman, we were bold enough to
give a dinner-party—or at least to invite four guests,
and trust to the *chef* of the Island Bakery to furnish a
suitable repast. But my trust was not strong, having
already had experience of his skill in small details;
and, notwithstanding the re-assuring fact that on one
occasion he had been curry-maker to the Duke of
Edinburgh, I awaited the coming of our little *fête*-day
with a certain amount of nervousness.

What a day of bustle it was! One would have
thought the Lord Mayor and the Corporation of
London were about to be entertained; and such was
the fuss, that I began to feel as if my party must
be growing every hour.

But if the commotion was great during the day, by
night it had swelled into a panic. Silver *entrée*-dishes
from the officers' mess were seen borne aloft on the
sable head of a Krooman, giving him quite a kingly
air; the more so, that he was the only one of the flying
messengers that had not lost his dignity and stately
pace. Presently a little dark-eyed St. Helena boy
rushes past with six champagne glasses in danger of
their fragile lives. Then, across their path, bursts, like

a meteor, a red-faced, white-aproned cook, with a stew-pan from the mess galley. A heap of plates and a cold tongue arrive from an opposite direction—canteen-ways. Viands approach from the four points of the compass, and gradually the bustle increases until a climax is reached at the door of the Bakery.

It appears that the baker had pressed into his service all willing hands, and had invaded every galley where talent was to be found, no doubt with a good purpose; but when I sniffed from afar this hurry-ing to and fro, I trembled for the consequences, and bethought me of a certain Scotch proverb, which says, " Too many cooks spoil the broth."

Our guests arrive punctually at 7 P.M., and I meekly ask a bright lad who is engaged as a waiter, if dinner is ready.

" Well, ma'am, the fish has been here half-an-hour, and everything has come except the soup," is the answer of this " enfant terrible."

Some idea of beginning at the wrong end flits across my mind, but I give it up, and resolve to wait for the soup.

All my housewifely sisters, who know what the pre-prandial ten minutes are to their nerves under ordinary circumstances, will pity me, left thus to the mercy of many cooks, and the agonies of fish waiting for soup. However, old Father Time, who carries off so many happy moments in his flight, is also kind enough to sweep away in their turn such trying moments as these, and at 7.15 the soup arrived.

It says much for the pleasant conversation of our friends, that no sooner had we sat down to table than I forgot my troubles, and was only sensible of everything being fairly good, and of a gradual increase of waiters as the dinner proceeded. Black waiters and white elbowed each other, and there was a good deal of knocking about in passing plates, accompanied by such stage whispers as " Get along, Jim." " Where's your 'ead ? " " *Ongtrys,* quick ! " On the whole, the party behind the chairs seemed to me to enjoy the proceedings thoroughly, and I only hope that the seated guests were as happy in their way.

The day after this domestic event was a very tiresome one. It was the 3rd of January, and had the latest news from the Cape been correct, in all probability the Mail would be in Clarence Bay before night. Under these circumstances, we were tied all the day to Garrison, with corded boxes and an unpleasant feeling of expectation. But no mail appeared. Our vigilance and our boxes were slacked, and as the hours passed on, we began to hatch plans for new excursions.

The turtle season had just commenced, but was, as yet, unproductive—indeed, the watchers had not been sent out ; and David and I were stimulated by an ambition to turn the first turtle of the season. So, armed with the Captain's permission, and a noose for the fins of our expected captives, we set out for Dead Man's Beach, accompanied by Brackley, about an hour after sunset on Saturday evening.

On this beach no regular watchers are stationed, and

it is the perquisite of the Kroomen to turn all the turtles that come ashore here ; a penalty of 5l. being inflicted on any person or persons found turning turtle without a licence.

Some yards above high-water mark, we came upon the little wooden hut used by the Kroomen in their watches, but as yet it was untenanted; and, while my husband and Brackley walked softly backwards and forwards along the water's edge, I spread out my rug here and lay watching the stars.

It was a lovely night. A pile of heavy, dark clouds lay in the east, and every now and again a single flake, detaching itself from the mass, would float overhead and sully for a moment the pure blue sky, that was glittering with hosts of stars. The moon was young, and as she gently glided towards the west, her crescent horn was reflected in a lake of silver on the dark waters of the ocean. When she had sunk to rest, then Venus lit up the sea. Never had I seen the Planet of Love so brilliant; she seemed to have cast off her own pale beauty, and glowed with the ruddier light of Mars. I watched her growing redder and redder as she sank lower and lower into the west, till at last Ocean, enamoured of her beauty, embraced her in his cold arms, and lo ! the sky was dark.

The noise of the water had been growing fainter and more distant in my ears, and I am not certain if the dreams that followed were altogether waking ones. I know I started unmistakeably when a voice whispered close to my ear, " It is ten o'clock, and there is no appearance of turtle. I think of taking you home, and then I shall walk out to South-west Bay to try our luck there."

Ten o'clock! then I must have been asleep. The air was blowing so softly that I felt loth to exchange my star-lit couch for the stuffy bedroom at Commodore's Cottage. But I could not spend the night here alone, and not having inclination for a longer walk I had no choice but to trot meekly home—serve my lord and master with a substantial supper, and go to bed; while he and Brackley again set out, after having securely locked me up with the dogs. The poor little beasts howled piteously at being left behind. The air was stifling, and I did not sleep nearly so well as I had done on Dead Man's Beach. I should have fared worse, however, had I attempted to accompany the hunting party to South-west Bay.

The tale of their misfortunes was comical. After a stiff walk of some three miles across the clinker, they reached the beach without mischance, and there, the first thing they saw was a dark moving object, a few yards in front of them.

Could it be a turtle? No, it was much too small. A wild cat? Yes—a cat at least, but hardly wild; not even "wildgewordene" (as Dr. Börgen very expressively terms the Ascension cats); for pussy advanced shyly towards David, and rubbed herself on his legs; mewing most pathetically all the while. Then she scampered off for a few paces, came back, and tried all her powers of coaxing to induce them to follow her, which they did, till they came to the door of the turtle-turners' hut, now empty. Here pussy scratched and whined, and, having finally led them to a water-butt, became very excited.

Poor suffering creature! she had been left behind by the men who usually dig limestone here when it is not the turtle season, and she was almost dying of thirst. David made all haste to get down water for her, which she drank most greedily for several minutes without lifting her head. Then he filled all the "panikins" he could find and placed them within her reach, so that she might be sure of a plentiful supply, until her careless owners should return to their work.

This done, they kept watch upon the beach for about an hour, but with no result, and then turned homewards, tired and ready for bed. But, alas! there were many slips between South-west Bay and Commodore's Cottage.

It was now a moonless night, with the sky overcast, and they somehow contrived to miss the steep narrow path which leads from the beach up to Waterloo Plains. After some ineffectual attempts to find the lost way, they succeeded in climbing up the face of the rock, at the cost of many bruises and much damage to clothing.

Nor were their troubles to end here; for, instead of the expected plain, they still found themselves among clinker—everywhere clinker, and wandered painfully over the rough ground for some hours. At last the clouds cleared away and showed the welcome sight of the Great Bear; and the wanderers, coming at the same time on signs of civilization in the shape of broken beer-bottles, were able to reach Garrison just before daylight—tired and turtleless.

There was only one excursion during these pleasant Christmas holidays that I was not able to take part in; but I need not omit to chronicle it on that account, because a paper of my husband's, written at the time, enables me to tell probably more of what was to be seen, than if I myself had been an eye-witness. His description of the abode of the birds interested me greatly, and, hoping it may be not without interest to my readers, I close this chapter with a long quotation :—

"I had long wished to visit Boatswain Bird Island, and fortunately an opportunity for doing so occurred during the last week of our stay in Ascension.

"The owners of a schooner that was frequently employed in conveying stores to Ascension had heard that on Boatswain Bird Island there were considerable supplies of guano, which they would be glad to purchase from the Admiralty, or to convey to England for a reasonable freight. To give the Admiralty a satisfactory answer to their questions on the subject, a visit to the island became necessary; and Captain Phillimore having offered me a place in the steam-launch, we set off one morning before sunrise to visit the abode of the birds.

"Ascension in the early morning certainly looks its best. A glorious sunrise we had, and the rosy

sunlight, illuminating the greys and yellows and reds
of the strange scenery, clothed it with a beauty seen at
no other time—redeemed it from the weird, and trans-
formed it into the beautiful.

"Much we enjoyed our sail as far as North-east
Point, but on rounding that, we encountered the long
heavy swell from which we had been sheltered in the
lee of the island, and our little craft began to knock
about in a way more lively than pleasant. The interest in the view of the island began to give way to an
uneasy sensation, which all sea-voyagers recognize as
a threatening of worse to come ; but before the alarming symptoms had time to develop fully, we were
safely anchored in the shelter of the friendly rock we
had come to visit.

"Only a furlong distant from the main island, it
rises 300 feet sheer out of 30 fathom water; its entire
surface being coloured pale yellow by a thin coating of
guano, formed by the birds which occupy every nook
and cranny of the steep sides, and cover nearly the
whole of the upper surface.

"The dingey was launched, and we pulled under a
projecting rock, from which swung a rope-ladder.
Climbing up, we reached a narrow ledge, whence, with
the assistance of a rope securely anchored above, a
good scramble brought us to the top. A perilous
job it must have been to climb the rock without this
assistance.

"But the climb was soon forgotten in the strange
interest of the place. On the sides we had encoun-

tered chiefly the smaller birds, black and white noddies, and I was much interested to find here in great
numbers our mysterious friend of Mars Bay, the
Sterna leucocapilla.

"An ornithological friend had positively assured me
that on his last visit to this rock, a year ago, not a
single specimen of this bird was to be found there, but
now the face of the rock was covered with them by
thousands.

" The beautiful white noddies, which I now saw for
the first time, flew about our heads with angry chirps
or little screams, and would then look reproach with
their mild black eyes as they flapped so close to us
that they could be caught by a quickly outstretched
hand.   No more graceful bird on the wing have I ever
seen than this delicate and pure white creature, with
its coal-black eyes, bill, and feet.

" Here and there we passed the nests of the beautiful
' Tropic ' or ' Boatswain ' birds, generally so shy and
unapproachable ; but woe betide the unfortunate hand
that disturbs them on their nests, unless with precaution and good protection.   A peck of that strong red
beak will go to the bone, and much amusement was
afforded by the repeated defeats of our first attempts to
secure some of the birds.

" But, if we were greeted with some noise by the
noddies as we ascended the sides, the sound was but as
a murmur to the infernal din that greeted our first
appearance on the top.   I had just put my foot on the
level, when a hideous scream behind me, followed by

acute pain in the calf of the leg, made me turn round in haste. I found the offender to be a great goggle-eyed yellow-billed gannet; and when I saw the long sharp bill and the wicked look of satisfaction at the wild work he was making with my trowsers, a sigh of thankfulness escaped me that I had not invaded Boat-swain Bird Island in a kilt. I revenged the damage by knocking him on the head.

"His next-door neighbour, however, did not seem to mind the treatment of his fellow in the least. There he sat glaring, goggling, and screaming with all his might, but not attempting to move, nor did the next, nor the next. There they were in rows, side by side, head to tail, in hundreds, a compact, screaming, goggling, quarrelsome mass.

"A few steps further, and what is this? A big, black, struggling lump with a red sack at one end— What can it be? After a few ineffectual struggles, a head develops out of the mass and rises up, a few struggles more and legs appear, and then with a flop, flop, flop, and half-a-dozen skips, a splendid frigate bird gets on the wing and floats away, the picture of ele-gance and grace. The great red sack, distended with water, hangs below his head like a grand beard, and sways gently with every turn of the graceful motion— the only speck of colour in his glossy black.

" Further on, there is more flop, flopping of wings, and this time it is not a single bird, but dozens and dozens of them, all swinging their sacks and struggling to rise, and making morning hideous with their screams.

More gannets, more frigate birds, more boatswain
birds at every step—colony by colony. Many refuse
to rise at all; others, having wheeled round three or
four times, alight again, but all this time they never
stop screaming—Pandemonium let loose!

"A considerable colony of Wide-awakes had established themselves here, doubtless, being warned by
previous experience, to escape plunder of their eggs by
the marines and St. Helena boys at the more accessible Fairs. The habits of these rock-dwelling Wide-
awakes were, however, in every respect identical with
those of their brethren on the plains—just as bold in
defence of their eggs, just as stupid in coming within
reach of capture.

" And now we set about our business; the Captain
to measure the guano, and I to make a collection of
the birds and eggs. There was no difficulty in getting
birds, but some trouble in getting a male and female
of each species, both young and old, as well as some
eggs of each. In this we at last succeeded, and with
great care and trouble conveyed the frail eggs down
the staircase of rope, and brought them in safety on
board the launch.

"Then up with anchor and back to Garrison; but
before we left, the engineer gave a blast of the steam-
whistle, and then what a row! A cloud of birds
darkened the sky, and we heard their frightened
screams till we were a mile off. We returned to
Garrison with high spirits and voracious appetites.
The birds and eggs were placed in the hands of our

kind and enthusiastic friend Mr. Unwin, who prepared the skins for transport to England.

" Our search for guano was not satisfactory; at least the quantity that could have been obtained seemed not sufficient to warrant the expense of the arrangements necessary for its shipment. There were only two or three inches of pure guano over most parts of the island, and in many places the surface was merely coated.

" How long must the great Guano Islands have been peopled by sea-fowl to yield the enormous supplies that have enriched England ? "

CHAPTER XXIV.

THE DEVIL'S RIDING SCHOOL.

THE sun has set on our last Ascension Sunday; but
a day or two yet remain for us to scramble among the
clinker, and we have determined on a visit to the
" Devil's Riding School."

When, from Green Mountain, we had looked down
upon all the little craters that are scattered over the
plains, we had longed to get to them, and this was our
first opportunity. At four o'clock on Monday morning
we were astir, and, having well broken our fast, were
ready within an hour to set out crater-climbing.

Dear old Jimmy Chivas was patiently waiting for
me, tied to the verandah gate ; Rover, and Brackley's
little terrier, Captain, were wild with excitement, and,
while we were getting ready, they kept running hither
and thither, kicking up dust in all directions, and
barking furiously ; no doubt to the annoyance of the

still slumbering Garrison. Fortunately Jimmy Chivas
had seen too much of the world's vanities to care to
join in these frolics, and started off sedately, after I
had comfortably seated myself on his poor old back,
and hung from the pommel of the saddle a leather
bag, containing some bottles of ginger-beer and designed to carry a return freight of clinker souvenirs.

The night had been tolerably cool, and, as we turned
southwards just before sunrise, we were met by a chilly
breeze that was perfectly delightful. For about a mile
we trudged along the Mars Bay road, seeing no life
and hearing no sound but the occasional cry of the
Wide-awake and the barking of the dogs, as they chased
in high glee over the clinker a wandering mule that
had inadvertently crossed our path. Soon we turned
off obliquely from the main road into a footpath, which
led us for about half a mile towards a little hill in the
south-east. So far there was no difficulty, as we had
received distinct verbal directions; but having unfortunately packed up our chart, and not finding here-
abouts a finger-post as we had expected, we now began
to feel an unpleasant sensation of doubt.

Here we were on a level plain, surrounded by five or
six great heaps, any of which might be the Devil's
Riding School, but which equally well might not; and
as these heaps were not mole-hills, the expedient of
climbing each in turn, until we hit upon the right one,
would be a fatiguing process. David, however, proposed to climb at least the little hill at the base of
which we were standing in indecision, hoping by

this means to gain a clearer notion of our whereabouts.

This he did, and then signalled for me to follow. I
thought from its description that this heap could not
be the one we were in search of—it was too red—but
I was glad of an excuse to take it by the way, and,
leaving Jimmy under care of Brackley, I began the
ascent, by no means such an easy matter as I had
imagined, for the slope was at an angle of 45°, and the
soil was so uncompacted that each footstep created a
miniature landslip.

When at last I reached the top, I found myself
standing on a narrow ridge, surrounding a great flat-
bottomed basin on all sides except the south-east,
where the wall was broken away. The ridge rose pre-
cipitously twenty or thirty feet from the basin, the
bottom of which was so level that one might have
played bowls over its whole extent; a valley on the
top of a mountain, and so shaped that we had no difficulty in recognising the " Horse-Shoe Crater."

Abutting from the south flank of this crater, another
of like form and colour towered twice as high above
it; and as we could not quite make up our minds about
the Riding School from our present position, David
started on a further voyage of discovery up this higher
hill.

Meanwhile I scrambled down again with many un-
dignified slips, as the treacherous scoriæ broke up
beneath my feet, and rolled in little fragments upon
the plain. Once I was seized with a terrible heart-

thumping as I heard a loud rumble, rumble behind me, while faster and faster the red stones came rattling down upon my heels. Suppose I have trod on some weak part of this great ash-heap, and it is going to bury me in revenge! flashed across my mind before I took courage to look behind and beheld that wicked little Rover slipping and scrambling down in my wake. With white silken paws daintily touching the loose red stones, he was borne onwards and downwards, amid a cloud of dust. He looked the picture of terror, poor little fellow! and as nothing would induce him to go on in front of me, I had no choice but to take him in my arms, and, thus laden, complete the descent.

Here at the base I waited until David should reach the top of the hill, up whose steep red side he was now clambering with hands and knees, hoping that from the summit he might be able to look down upon the Riding School crater, and direct me to it. Fortunately he could do so, and his dark figure stood out so strongly against the light background of the sky, that, when he perched himself on the edge of the unbroken crater-cup, I was able to see distinctly the direction he indicated with his alpenstock.

Thus guided, I rode due east across the plain until I reached a light-coloured hill, lower than some of the surrounding ones, but apparently of considerable circumference. Here Jimmy and I waited until my husband overtook us, and then I was glad to learn that we were now at the Devil's Riding School.

The lithological formation of the slope which rose above us was certainly a contrast to that of the last crater I had climbed; for here we found firm footing on the large grey stones well-compacted together, and dotted all over with lichens, while meek little tufts of coarse grass peeped from the crevices. We reached the top without fatigue, and then found that we had been so fortunate as to gain at once the highest part of the circular summit, which was very irregular, and in some places did not rise many feet above the depressed centre. I say fortunate, because the greater elevation enabled us to obtain a clearer idea of the whole than we should have done, had we come at once upon the level.

We looked down upon a bit of landscape, so extraordinary in its marked contrast to the careless, irregular beauties of a natural hill, that to a non-scientific mind it almost suggested the supernatural.

The circus appeared to us about a mile and a-half in circumference, and, from our standing point, it seemed to be perfectly level. In the centre there is a reddish area of considerable extent, surrounded by a narrow rim of a very light sandy colour; then a dark ring, in turn surrounded by a broad white circle; while great masses of rough grey stone form an almost unbroken fence round these fairy rings. There is indeed one narrow gap—the evidence of an outflowing towards the south-east—but, viewed from a little distance, this hardly breaks the circle; and in some places I should imagine the fence to be as high as thirty feet. At such a height we now stood, and viewed and wondered.

What is it that we see? What has once been here? A vomiting of mud? A waterspout? A lake? It was impossible for us to say, nor could we tell whether this too was a hill of sudden upheaval, like the little volcanic chimneys around, or if it were some child of slower growth.

On the Admiralty chart the "Devil's Riding School" is marked, "Crater of an old volcano," but Darwin, in his "Volcanic Islands," rejects this description as incorrect, and contends that it is no volcano at all. He says, "The hill marked in the map 'Crater of an old volcano,' has no claims to this appellation which I could discover, except in being surmounted by a circular, very shallow, saucer-like summit, nearly half a mile in diameter. This hollow has been nearly filled up with many successive sheets of ashes and scoriæ of different colours, and slightly consolidated. Each successive saucer-shaped layer crops out all round the margin, forming so many rings of various colours, and giving to the hill a fantastic appearance. The outer ring is broad and of a white colour, hence it resembles a course round which horses have been exercised, and has received the name of the Devil's Riding School, by which it is most generally known. These successive layers of ashes must have fallen over the whole surrounding country, but they have all been blown away, except in this one hollow, in which pro-

bably moisture accumulated, either during an extraordinary year, when rain fell, or during the storms
often accompanying volcanic eruptions."

But whence came the hollow in the beginning, and
whence the rocks that form its flanks? Before the
ashes and scoriæ * fell, there must have been, in the
first place, a hill with a hollow summit, rather a novelty
where there is no volcanic origin, and, as there is certainly no *other* hill on the island except those formed
by the accumulations of craters, an ignorant observer,
like myself, would be inclined to ascribe a like origin
to the Devil's Riding School.   This especially, as the
form—circular, with a depressed summit—is similar to
that of the surrounding hills.

To be sure the formation and colour are different,
and this tells a story intelligible enough to the skilled
geologist, but a novice in the language of stones is lost
when a fresh leaf is turned ; and, finding the new page
unlike the old one, he is glad, after hopeless puzzling,
to throw away his own bewildered ideas, and rest his
mind on the master's teaching without question.

But I had not read Darwin's " Volcanic Islands "
when I visited the Devil's Riding School, and it was
with a feeling of baffled curiosity that I descended
from the lip into the saucer.   As we walked across it,
we found this saucer to be by no means so level as we

---

* With regard to these strata of ashes and scoriæ, a German writer,
quoting from Ehrenberg, says, " Aus organisirt gewesener Substanz er
findet einige Kieselschalige Susswasser-Infusorien und nicht weniger
als 25 verschiedene arten Kieseliger Gewebe von Pflanzen, hauptsächlich
von Gräsern. "

had supposed, and that the outcrop of the various strata occurred at very different depths. Dotted over many parts of the circus too (particularly on the north side), there were little cones of one and two feet in height, and, on having their heads knocked off, these displayed tiny central chimneys—looking as if they were meant to represent Ascension in miniature.

It was all very interesting and curious, and we would gladly have spent a much longer time poking about among the ashes, had not the sun warned us that by-and-by he would make himself very disagreeable on the plains. For our descent we chose the north-east slope, and found it of a character altogether different from that of the south-west, by which we had come up.

Here the stones were mostly loose, and curious pieces, like saucepan lids, lay strewn about in every direction—very metallic in sound when struck with the hammer, or against each other, and very hard and sharp for the feet, making walking difficult. In some places they were securely fixed, and stood out at right angles to the slope of the hill.

I find from Darwin that these saucepan-lids are " veins," which intersect the trachyte in the most complicated manner.

He says, " They are best seen on the flanks of the ' Crater of an old volcano.' . . . The veins vary much, and suddenly, from the tenth of an inch to one inch in thickness ; they often thin out, not only on their edges, but in their central parts, thus leaving

T

round, irregular apertures; their surfaces are rugged.
They are inclined at every possible angle with the
horizon, or are horizontal; they are generally curvi-
linear, and often interbranch one with another. From
their hardness they withstand weathering, and projecting two or three feet above the ground, they occasionally extend some yards in length. . . . . Their
fragments, which are strewed on the ground, clatter
like pieces of iron when knocked against each other.
They often assume the most singular forms; I saw a
pedestal of the earthy trachyte covered by a hemi-
spherical portion of a vein, like a great umbrella,
sufficiently large to shelter two persons."

Curiously enough, we descended at the very spot
where this umbrella stands, actually alighting on the
top of it; unless, indeed, there be more than one such
phenomenon here, which is unlikely.

After our dusty climb, we were glad to return to
Jimmy Chivas, with his burden of ginger-beer. Hot
and dusty we longed for a draught, but alas! we had
forgotten a glass, and could not quench our thirst so
pleasantly as might have been desired. However, with
patience and many chokings, we cured the worst of it,
pouring what remained of the beer into one of the
stony saucepan-lids for the dogs; but they'd none
of it.

Then we filled the empty bag with stony treasures
from the hill, and left our quota of broken bottles to
glisten on the clinker. These broken bottles strewn
everywhere, over plains, roads and hill-sides, form a

very characteristic feature in the Ascension landscape.
I wonder, will their " fossils " puzzle future geologists,
and lead them to mistake Ascension for the *Bass* Rock
of this age !

We paid for our lengthened stay at the Riding
School by a hot, tiring journey home.   The sun had
long since o'ertopped Green Mountain and its clouds,
and was now shining in an unveiled sky, dazzling
our eyes and making my head ache, so that I
greeted with thankfulness the welcome sight of
Garrison, which we reached a little before noon.   We
were home again from our last excursion on the
clinker, tired mentally as well as physically ; for the
effort of trying to understand the rough road was
quite as fatiguing as the exertion of walking over it.
We felt that there was very much even in this little
spot of earth that puzzled us, and six months' residence on a barren rock of only twenty-eight miles in
circumference, by no means justified our claiming intimate acquaintance with it.

On the evening of the 8th, my husband made his
last observation for time, in the disused Magnetical
Observatory across the square from Commodore's
Cottage.   Here he was invaded at 10 o'clock by the
Corporal on guard, who demanded " Lights out," and
was much surprised to find that the astronomer was
not yet quite " packed up," but continued to trans-
gress rules till the last moment.

These last Ascension days and nights were cloudless
—just such days and nights as had greeted our arrival,

so that all the necessary time-observations were comfortably secured, and everything was well ended.
Finis coronat opus.

Wednesday, 9th.—No steamer, and we began to
wonder whether Ascension had been forgotten!
Thursday morning—still waiting; but, while I was
sitting quietly with my needlework at 4 P.M., the white
flag and ball were suddenly hoisted on Cross Hill:
Mail in sight!

My needle was left half undrawn, and all at once I
felt in a bustle, without exactly knowing why, for we
had been ready a long while.

Within an hour of signalling, the *Warwick Castle*
anchored in Clarence Bay, and, now that she was
actually here, I almost wished she had not come. I
did not like saying good-bye, nor did I enjoy the
prospect of a fortnight's voyage; and, when the
Captain told us that we might still have some hours on
shore, I felt as if a respite had been granted me.
These hours were occupied with last words and adieux,
not the least touching of which was my parting with
Rover, who was left, however, with a kind mistress,
and looked by no means inconsolable.

At sunset we embarked, and before steam was up
had time to arrange some comforts in our cabin,
and to read the home letters which had come by the
mail. These were delightful, telling as they did, that
the Mars Expedition was considered, by those well-
qualified to judge, to have been energetically and satisfactorily carried out—a pleasant thought to be sea-sick

upon, and I felt brave enough for worse trials than the Bay of Biscay!

What a lovely evening it was, and how gloriously the crescent moon and Venus shone over the water and silvered the grim red and brown outlines of the land!

"For the last time," I said, to myself, as the evening bugle sounded; and, before its echoes had died away, our "Six Months in Ascension" was a thing of the past.

CHAPTER XXV.

HOMEWARD BOUND.

Cheerful neighbours.—Sunday at sea.—Sitting on deck.—Teneriffe again.—Cooler latitudes.—Rousing the inhabitants.—Funchal by moonlight.—An unexpected meeting.—American Astronomers.— Victims of *curio* vendors.—The Bay of Biscay.—A gale in Channel.—Home.

AND now followed a repetition of sea-discomfort— nausea and stuffiness; but on this occasion it was short lived, for after a couple of days the captain kindly arranged that our cabin, near the screw, should be changed for one on deck, far forward.

Here we caught the welcome current of air caused by the ship's motion, and for the first time in my sea experience I awoke in the morning refreshed. I awoke too, with the sweet sense of home upon me, and pleasant recollections of a certain dear old farmhouse; for my dreams had been mingled with the bleating of sheep, the cackling of geese and the crowing of cocks. Poor things, there was little chance of their being led out to green pastures, or participating ever again in the varied pleasures of a fascinating dung-hill. Pri- soners they were, under sentence of death, but they crowed lustily, and " ba'a ba'aed " sweetly, nevertheless ; and, grateful for the cheer their good spirits gave to me, I would hope that they dreamt not of to-morrow. I know not whether to ascribe it to the inspiriting influence of these, my feathered and four-footed neighbours, but certain it is, that on the third day of voyaging I could see things straight, and had sensations of pleasure in the prospect of dinner.

It was a welcome surprise; and on Sunday I had the privilege of being able to attend divine service at sea for the first time. Always and under all circumstances are the prayers and collects of our English prayer book touching and beautiful, but they appeared to me especially so on that Sunday.

" O Eternal Lord God, who alone spreadest out the heavens, and rulest the raging of the sea ; who hast compassed the waters with bounds until day and night come to an end,"—words so fitting that it seemed as if the murmuring sound of the waters, lapping against the ship's sides, was His gentle answer to our prayer—" That we may return in safety to enjoy the blessings of the land, with the fruits of our labours."

We had very few passengers on board and only three of the number were ladies, yet this was to me an *almost* enjoyable voyage ; for during the greater part of the time I could sit on deck all the day, watching the dolphins and flying-fish at play in the unruffled sea ; and at night, what could be more beautiful than

the water lit up by the balls and streams of phos-
phorescent light, which the ship seemed to cleave
in her course, throwing them off to right and left of
her, and leaving a golden river in her wake ?

Again we are approaching Teneriffe ; and I really
ought to pass it by without comment, for it is long,
long ago since Humboldt wrote, " From every traveller
beginning the narrative of his adventures by a description of Madeira and Teneriffe, there remains now
scarce anything untold."

Great pens scorn a scribbled page ; but mine must
not skip (I leave my reader to do that), and I tread
again the old track in my own footsteps, which are so
slight that, methinks, a re-impression is their only
chance of notice.

Again we are at Teneriffe ; but it is not the Teneriffe of six months ago.   Then the Great Peak was in
sunny summer garb, trimmed with bright floating
ribbons ; now we only see a snow-covered head in
the heavens, and the horizon below filled with the
shadowy outline of a giant wrapped in a mantle of grey
cloud.

It was here that we sniffed the first breath of
" caller " air.   No more tepid water to drink ; no
more folding of the hands to rest ; and, when we
arrived at Madeira two days later, the poor little diving
boys showed blue lips, and shivered as they chattered
in their scanty dripping garments.

Having had the welcome news that time would per-

mit of our going ashore at Funchal for a few hours, we
were very impatient for the custom-house and health-
officers to come off while daylight lasted. But, according to their usual wont, they did not hurry, and the
Captain, getting impatient as well as we, blew repeated
whistles and roused splendid echoes among the rocks
with his monster steam-horn.

At last two lazy-looking Portuguese officials came
alongside, whereupon one of them called out in his
distinct foreign English, "Sorry, Captain, but I think
somebody on board has got a very sore throat."

"Sore throat!" said the Captain. "No; why
should you think so?"

"Because very hoarse cries come from your ship
and disturb us," was the answer, with a good-humoured
laugh; and the joke was enjoyed by all on board, including the Captain himself, whose noisy method of
commanding attention in lazy ports was well known to
all the ship.

After some lively bargaining from the gangway, a
party of us got into a boat and rowed ashore—unfortunately just as the sun was setting; but soon a
glorious moon took his place, and, seen by her light,
the group of dusky men, who almost seized upon us as
we set foot on the beach, clamouring to be taken as
guides, was a striking and strong-coloured picture. In
the rising background glimmered pretty white houses;
and there was quite light enough for us to admire the
glorious masses of red and purple Bougainvillea, which

over-ran the garden-walls and terraces, making a perfect Eden to our flower-starved eyes.

We were in search of no lions, and wanted no guides
—simply a stroll on *terra firma*, and a peep down the
narrow streets. Everywhere, as we passed along, dark
eyes questioned us with curious glances ; and every
now and then we encountered some portly matron *en
route* for an evening party—not in a fly, nor in a sedan
chair, but in a " thing," partaking somewhat of the
nature of both : a sort of wheelless couch, or rather
sleigh, hung with curtains, and generally dragged by a
couple of small bullocks.

Everything was dragged up and down these steep,
slippery streets, which are mostly paved with small
pebbles, edge upward, and worn smooth by constant
friction. This mode of locomotion seems easy and
pleasant, and certainly gives an air of repose and
leisurely dignity to the scene. Life seemed to pass
slowly on this pretty island, and, in the feverish
bustle of London life, one's thoughts turn to it with a
sense of rest.

By the merest chance, my husband heard that an
American astronomical party was here for the purpose
of making longitude determinations ; so, leaving me in
the hands of some of our fellow-passengers, he went
off to find out these kindred spirits. This he had no
difficulty in doing, and they welcomed him most kindly
and courteously ; one of the party taking the trouble
to bring a chronometer on board to compare it with

ours, thus making a strong rivet in our longitude run from Ascension to England.

Meanwhile I searched several of the shops, hoping to buy some of the cotton embroidery which at home we value so much for trimmings, but I could find none so good nor so cheap as what I had seen in London ; so I confined my purchases to some sprays of feather flowers and a few fancy baskets, all of which I lost at Plymouth!

The meeting of our party at the boat was most amusing. Everybody had been victimised in some way by *curio* vendors except David, who was empty-handed, and kept an anxious eye on his American friend with the chronometer, when, at the last, wicker-chairs and other goods were being thrown helter-skelter into the boat. We were a perfect floating bazaar, with our baskets, chairs, poodles, embroidered eggs, walking-sticks, mats (sewn with soap-berries and Job's tears), and such like.

Some young officers who were of our party had decked themselves with peasant caps, made of soft black cloth, fitting tight to the forehead, and terminating in a long upright peak, which gave them quite an air of Mephistopheles in the moonlight.

These hours ashore made a charming variety in the monotony of sea-life and were perfectly enjoyable, if one could but have forgotten the Bay of Biscay lying between this peaceful haven and Plymouth Harbour. After Madeira the weather became quite cold, and all summer attire rapidly vanished into trunks, giving place to long-neglected serge gowns and fur overcoats; still wind and sea were gentle, and it was only when within two days of home that the face of sky and water changed for the worse.

From that time things continued to grow worse and worse, until the last night, when it blew a regular gale in the Channel. The sea ran very high, and every lurch of the ship sent a flood of water sweeping over the decks. Heavy boxes were skipping about the cabin, where I lay packed into my berth with numerous pillows, sick and excited. It was especially miserable during the night, which I am sure is the longest on record. All through its weary dark hours the Captain was on the bridge, and constant signalling went on to the engine-room; but so thick were fog and blinding rain and so strong the current, that we were able to make neither the Lizard nor the Eddystone Light-house, and daylight found us several miles up Channel, with Plymouth behind us.

But a few turns of the screw brought us back again, and at noon on the 24th of January we once more set foot on England. Dear old England! Still warm to our hearts, in spite of her cold reception of wind, rain, and snow-showers.

Land—England—a fireside! It is only when these are again in sweet possession that one thoroughly enjoys travelling. Then, when all bodily fatigue and discomfort are forgotten, and there still remain in the

mind pleasant pictures of strange life and scenes, and
in the heart the softening influence of a wider knowledge of men and things—it is then that the delights
of travelling are fully known, for it is then, and only
then, that one is able to enjoy to the full the pleasures
and privileges of Home.

THE END.